Data measured in the form of angles or two-dimensional orientations are to be found almost everywhere throughout science. They commonly arise in biology, geography, geology geophysics, medicine, meteorology and oceanography, as well as many other areas. Typical examples of such data include the departure directions of birds from points of release, orientations of fracture planes, the directional movement of animals in response to stimuli, directions of wind and ocean currents, and biorhythms.

Statistical methods for handling such data have developed rapidly in the last 20 years, with emphasis on problems of data display, and the correlation, regression and analysis of data with temporal or spatial structure. In addition, some of the exciting modern developments in general statistical methodology, particularly nonparametric smoothing methods and bootstrap-based methods, have contributed significantly to the data analyst's ability to make progress with problems that have been relatively intractable.

This book provides a unified and up-to-date account of techniques for handling circular data. Extensive use of clear diagrams, together with numerous tables, make this an essential handbook for the professional data analyst.

T0192226

Statistical Analysis of Circular Data

N.I. FISHER

Division of Mathematics & Statistics, CSIRO Australia

CAMBRIDGE
UNIVERSITY PRESS

CAMBRIDGE UNIVERSITY PRESS
Cambridge, New York, Melbourne, Madrid, Cape Town, Singapore,
São Paulo, Delhi, Dubai, Tokyo, Mexico City

Cambridge University Press
The Edinburgh Building, Cambridge CB2 8RU, UK

Published in the United States of America by
Cambridge University Press, New York

www.cambridge.org
Information on this title: www.cambridge.org/9780521568906

First published 1993
Reprinted 1995
First paperback edition 1995

A catalogue record for this publication is available from the British Library

Library of Congress Cataloguing in Publication Data

Fisher, N. I.
Statistical analysis of circular data / N.I. Fisher.
    p.   cm.
Includes bibliographical references and index.
ISBN 0 521 35018 2
1. Mathematical statistics. 2. Circular data. I. Title.
QA276.F488   1993
519.5–dc20   92-23165   CIP

ISBN 978-0-521-35018-1 Hardback
ISBN 978-0-521-56890-6 Paperback

STATISTICAL ANALYSIS OF CIRCULAR DATA

To Judy, Rachel, Stephen and Anna

Contents

Preface to first paperback edition

This edition incorporates a number of corrections to the text and figures of the first (1993) edition. I am grateful to all those who have drawn my attention to errors or misprints.

Nicholas Fisher
June 1995

Preface

1 The purpose of the book

Data measured in the form of angles or two-dimensional orientations are to be found almost everywhere throughout Science. They arise commonly in Biology, Geography, Geology, Geophysics, Medicine, Meteorology and Oceanography, and in many other areas. Typical examples include departure directions of birds or animals from points of release, orientations of fracture planes and linear geographical features, directional movement of animals in response to stimuli, wind and ocean current directions, circadian and other biorhythms, times of day of accident occurrences, and so on.

The last 20 years, and more particularly the last 10 years, have seen a vigorous development of statistical methods for analysing such data, with emphasis on problems of data display, correlation, regression, and analysis of data with temporal or spatial structure. In addition, some of the exciting modern developments in general statistical methodology, particularly nonparametric smoothing methods and bootstrap-based methods, have contributed significantly to the data analyst's ability to make progress with problems which have been relatively intractable. The subject has now reached a point of development at which it seems appropriate to provide a unified and up-to-date account of this material for practical use. In this respect, the present book is a companion volume to *Statistical Analysis of Spherical Data* (Fisher, Lewis & Embleton 1987) which was concerned with *three*-dimensional unit vectors or orientations, although developments in nonparametric smoothing and bootstrap methods over the last four years have meant that rather more effective use of them has been possible in this book.

As with the spherical data book, this book is directed at several categories of reader:

(1) to the working scientist collecting and interpreting circular data;
(2) to university students, whose course-work, laboratory experiments or research work requires an understanding of circular data;
(3) to statistical research workers, as a guide to the status of the subject.

Similarly, priority has been given to the first of these groups, so that the book is primarily a manual for the working scientist. The emphasis has been on applications, with the data examples being drawn from diverse scientific fields. Generally, I have sought to provide a logical approach to displaying, analysing and interpreting circular data. In any given situation, an appropriate statistical method is described up to the point of actual implementation, and reference given to the supporting statistical theory. Not uncommonly, several essentially equivalent methods are available for a particular problem, in which case a single choice has been made in keeping with related problems, and the other methods have been referenced. Tables are provided for commonly-required distributions or specialised purposes.

For the most part, the level of mathematics required to use the book would be that acquired in a first-year university mathematics course. The principal exception to this is in the programming of bootstrap techniques into a computer. My experience of introducing these methods to scientists over the last few years suggests that the easiest way for a scientist who is not a skilled computer user to implement the methods is to find a colleague or an assistant who is skilled in this way. The benefits of doing so will more than justify the search.

2 Survey of contents

Chapter 1 provides a historical overview of the subject. In Chapter 2, we address the basic issue of displaying a single sample of circular data so as to highlight its main features, and introduce a number of basic sample summary quantities. Chapter 3 describes a number of probability distributions available as potential models for circular data. Chapter 4 gives an account of exploratory analysis of a sample of circular data, followed by methods for summarising the information in a single sample and evaluating statistical hypotheses about the population from which the sample was drawn. Chapter 5 considers methods for comparing two or more samples of data and, where appropriate, combining the information in the samples to form pooled estimates of parameters of interest. Chapter 6 is concerned with problems of association between a circular random variable and another random variable, and of modelling a circular response variable in terms of a number of explanatory variables (that is, regression models with a circular response variable). Chapter 7 deals with problems in which the circular random variable under study has an associated temporal or spatial dimension. Finally, two important general classes of methods – bootstrap methods and randomisation methods – are explained in Chapter 8, which is a reference chapter for the earlier ones.

3 How to use the book

Each of the main chapters on data analysis (Chapters 3–7), and each main section within these chapters, begins with ways of exploring the data set to hand, proceeds to general methods of analysis based on relatively few assumptions, and finally (in some instances) to methods based on parametric models. It is recommended that the non-statistical user follow this path as well. In many instances (e.g. calculating a point estimate and confidence interval for a mean direction), the general methods will suffice.

4 Notation, terminology and conventions

Unlike spherical data, there are very few methods for recording circular measurements. Whilst they may be collected in different units, such as times of day, converting them to degrees or radians is trivial. Sometimes we shall work with data in the range $(0°, 360°)$ and at other times in the range $(-180°, 180°)$ (or the corresponding intervals in radians). Apart from variations of unit, the principal variation likely to be encountered is the use of polar coordinates $x = \cos\theta$, $y = \sin\theta$ to represent the direction θ as a point on the unit circle. When plotting circular measurements, it should be borne in mind that, mathematically, angles are measured anticlockwise from the point (1,0) on the unit circle. To have the directions plotted so that compass directions appear in their customary positions, the preliminary transformation $\phi = 90° - \theta$ is required.

A key distinction used in this book is between *vectorial data* and *axial data*. A vector is a directed line, for example the departure direction of a homing pigeon, or the direction of the nearest public house. Vectorial data will be displayed by plotting points on the unit circle, using an arrow head as the plotting symbol. An axis is an undirected line, such as a fracture in a rock exposure, where there is no reason to distinguish one end of the line from the other. More generally, one can have p-axial data, where for $p = 6$ say, the directions $10°, 70°, 130°, 190°, 250°$ and $310°$ would all be regarded as equivalent. In this book, axial data will be plotted with a circular blob at each end of each axis (so that a sample of n axes will be plotted as $2n$ blobs around the unit circle). Axial data present few additional problems for statistical analysis, as they can be converted to vectors (by doubling them and then reducing them *modulo* $360°$), analysed as such, and the results back-transformed. Whilst the term 'orientation' can refer to either a vector or an axis, the term "trend" will generally be taken to refer to an axis.

Another frequently encountered pair of words will be 'unimodal' and 'multimodal', the former referring (in the context of a single sample of data) to the fact that there is a single cluster evident in the data, and the latter to the existence of two or more clusters. A sample of axial data exhibiting a single cluster of axes will sometimes be referred to as a *bipolar* data set, because the method for plotting axial data shows two modal groups; the phrase is more widely known in the context of spherical data.

5 Acknowledgements

Many people have provided invaluable assistance in the production of this book.

Judy Lain and Kathy Cocks typed substantial portions of the first draft. Michael Buckley and Glenn Stone contributed numerous pieces of advice about the intricacies of TeX. CSIRO librarians at the National Measurement Laboratory and the Minerals Research Laboratories tracked down and secured several obscure references. Simon Mitton (Cambridge University Press) was most understanding about retreating deadlines, and also obtained copies of figures from several publications which were inaccessible in Australia.

Much of the basic work was done during visits to the University of Rochester (NY) and Oxford University in 1987. Numerous data sets were supplied to me upon request (see Appendix B for specific acknowledgements); Garry Tee, Michael Stephens and Geof Watson contributed several references and pieces of historical information; and Judy Fisher and Peter Hall provided critical comment on drafts of some of the chapters.

The contributions of the various institutions and of all these people are gratefully acknowledged.

Two friends require special mention. Steve Davies did a lot of careful software development, using S and S-PLUS, to turn my basic set of ill-matched graphs into the clean, unified set which appears in this book. Toby Lewis made a very careful reading of the first draft; his detailed and constructive comments have led to major improvement in the presentation, as well as the correction of numerous errors. Residual faults are my responsibility.

The Greengate   *Nicholas Fisher*
Killara
February 24, 1992.

1

Introduction

Circular data analysis is a curious byway of Statistics, sitting as it does somewhere between the analysis of linear data and the analysis of spherical data.

For one thing, various aspects of the subject seem trivial in comparison. A sample of linear data can be so disposed as a collection of clusters, long tails, inliers and outliers over an indefinitely large interval of the real line that no single display shows all features. A sample of spherical data can be so dispersed over the surface of the unit sphere as modal groups, transitional modal-girdle patterns and outliers that it can really only be viewed satisfactorily using a rotating three-dimensional display. By contrast, circular data cannot escape very far from each other (the notion of an *outlier* is of little consequence in most practical situations) and certainly cannot hide from view. So, many of the basic problems addressed in the exploratory phase of linear or spherical data analysis seem benign or non-existent for circular data.

For another, the problem of extracting practical information from circular data would appear to be tantalisingly close to the same problem for linear data, especially for concentrated data sets (i.e. data sets contained in a small arc of the circle). Approximate linearity of a small arc would seem to justify application of linear methods and so make special treatment of circular data largely unnecessary.

Both these observations can readily be seen to be facile. A basic problem occurring commonly with geological orientation data is to decide how many distinct modal groups are manifested in a sample of data. The very finiteness of the circle creates new problems as readily as it solves others. And while linear approximations may solve *ad hoc* data analysis problems they are not suitable for routine data processing. The notorious crossover problem which arose in automating the summarisation of wind direction data (see Chapter 3) occurred for just this reason.

Graphs featuring circular data can be found in writings dating from the end of the First Millennium. Funkhouser (1936) (see also Beniger & Robyn,

1978) discussed an early example of such a graph, from the 10th or 11th century:

> The main body of the text contains the two books of the commentary of MACROBIUS [AMBROSIUS THEODOSIUS MACROBIUS, Roman grammarian and philosopher, flourished about 400 A.D.] on CICERO's *In Somnium Scipionis* in which MACROBIUS reviews the status of physics and astronomy of his day. There is an appendix *De cursu per zodiacum*, possibly added by the unknown transcriber of the tenth century, which is a short description of the movements of the planets through the zodiac. The graph [Figure 1.1] is intended as an illustration for this compendium. The whole work seems to have been compiled as a text for use in monastery schools.
>
>   The graph apparently was meant to represent a plot of the inclinations of the planetary orbits as a function of the time.

A fascinating account of the discovery of the directive properties of a magnet in the earth's field, and of the claims and counter claims of priority in relation to observation of magnetic declination, can be found in Chapman & Bartels (1940, Volume 2 Chapter 26). The earliest recorded measurements of magnetic declination appear to date from about 1510. By 1701, Sir Edmund Halley was able to publish a map which showed lines of magnetic declination for all the navigated waters of the world; a portion of this is displayed in Figure 1.2.

The early roots of circular data *analysis* reach back at least as far as the mid-18th Century. In 1767, the Reverend John Mitchell FRS analysed angular separations between stars, and found that the number of close pairs was too large to be consistent with a hypothesis that the directions of the stars were uniformly distributed.† The paper continued:

> And the natural conclusion from hence is, that it is highly probable in particular, and next to a certainty in general, that such double stars, &c. as appear to consist of 2 or more stars placed near together, do really consist of stars placed near together, and under the influence of some general law, whenever the probability is very great, that there would not have been any such stars so near together, if all those that are not less bright than themselves had been scattered at random through the whole heavens. (Mitchell 1767, p. 432.)

That is, Mitchell inferred that such pairs of stars were physically linked by gravitational attraction.

---

† An even earlier reference to the problem of testing uniformity is to D. Bernoulli in 1734: see Watson (1983, page 40).

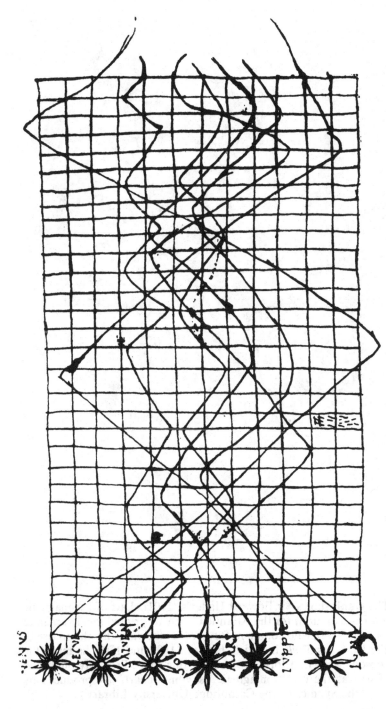

Fig. 1.1 This graph is believed to show a plot of the inclinations of the planetary orbits as a function of time. Funkhouser (1936) comments: 'For this purpose the zone of the zodiac was represented on a plane with a horizontal line divided into thirty parts as the time or longitudinal axis. The vertical axis designates the width of the zodiac. The horizontal scale appears to have been chosen for each planet individually for the periods cannot be reconciled. The accompanying text refers only to the amplitudes. The curves are apparently not related in time.' (Reproduced by permission of the Syndics of the Cambridge University Library.)

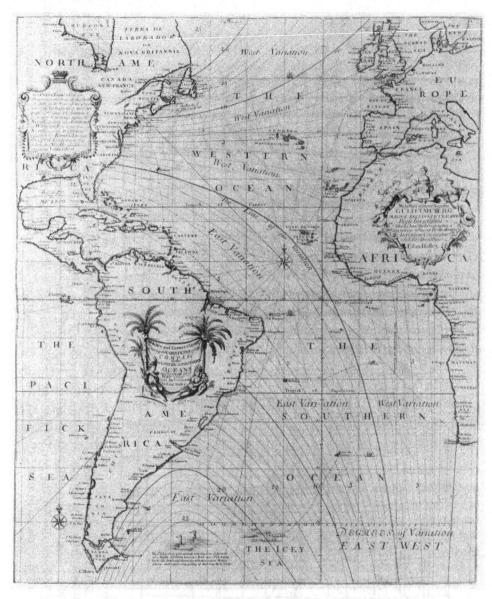

Fig. 1.2 The graph shows Sir Edward Halley's (1701) chart of the magnetic 'variation' (or declination) for the Atlantic Ocean, as printed in Chapman & Bartels 1940, Plate 38), who comment further: '... showing also the course of his ship, the *Paramour Pink*, in which, between 1697 and 1701, he made the magnetic observations that form the basis of this chart. Probably published in 1701. This was the first isomagnetic (isogonic) chart.' (Reproduced by permission of the Syndics of the Cambridge University Library.)

Half a century later, John Playfair pointed out the need to analyse directional data differently from ordinary data. In a short historical note, Cox (1990) drew attention to '... an aside on p. 230 in a Note (pp. 209–38) on "Elevation and Inflexion of the Strata" ':

> For the sake of those who would deduce the medium bearing of any body of strata from a number of observations, it may be proper to take notice, that the true average is not to be found by simply taking an arithmetical mean among all the observations. A more exact way is to work by the traverse table, as in keeping a ship's reckoning, (supposing the distance run to be always unity), and to compute from the observed bearings the amount of all the southing or northing, and also of all the easting or westing. The sum of all the latter, divided by the sum of all the former, is the tangent of the angle which the general direction of the strata makes with the meridian. (Playfair 1802)

As Cox observes, 'John Playfair recommended the use of the resultant vector method of averaging directions as early as 1802 ... So far as I am aware, this is the first mention of this idea in a geological context.'

Florence Nightingale had been a chief nurse in the British Army during the Crimean War, and had learned that thousands of lives could be saved simply by improving the sanitary conditions in military hospitals and barracks. In a very informative article, Cohen (1984) noted that

> At a time when the collection and analysis of social statistics was still uncommon Nightingale recognized that reliable data on the incidence of preventable deaths in the military made compelling arguments for reform. Thus in addition to advancing the cause of medical reform itself she helped to pioneer the revolutionary notion that social phenomena could be objectively measured and subjected to mathematical analysis.

In an 800-page privately printed book, Nightingale (1858) used a variety of methods to argue her case. One of these was a new graphical device which she called the coxcomb (because of its various colours) and which is now commonly referred to as a rose diagram†: an example is shown in Figure 1.3. This and other graphs were used to emphasise the effect of improved sanitation.

Since these early examples of circular data analysis in Astronomy and Medicine, applications have appeared in numerous fields – some largely untouched by statisticians – in subject areas as diverse as Biology, Geography, Geology, Metereology and Oceanography, of which the principal ones have been Biology and Geology.

---

† This differs from the nonstatistical sense of the term 'rose diagram', or 'wind rose', as used by mariners down the centuries to refer to the 16-point compass which ornamented their maps.

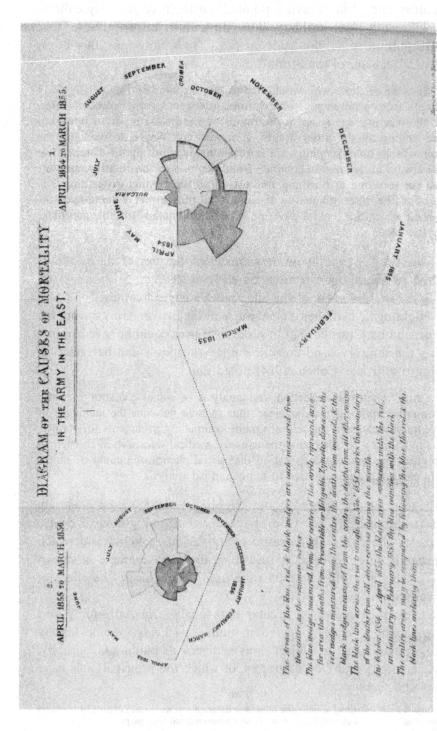

Fig. 1.3 The graph shows one example of how Florence Nightingale used novel methods of data display to argue her case for the efficacy of improved sanitation in hospitals. The two 'coxcombs' show the effect of her measures from one year to the next, and were originally printed in colour. As Nightingale described it, 'The Areas of of the blue, red and black wedges are each measured from the centre as the common vertex. The blue wedges measured from the centre of the circle represent area for area the deaths from Preventible or Mitigable Zymotic diseases [contagious diseases such as cholera or typhoid]; the red wedges measured from the centre the deaths from wounds; and the black wedges measured from the centre the deaths from all other causes ...' (Reproduced by permission of the British Library. 8831 bb 36 opp Pg. 315.)

Batschelet (1981) devoted a whole book to studying methods for analysing circular data in biological contexts; lengthy chapters in other books (Zar 1984, Upton & Fingleton 1989, Cabrera *et al.* 1991) explored the same area. Typical of the problems of interest to biological scientists are those of bird navigation and of general orientations selected by particular creatures in response to experimental variation of their natural habitat (or of parts of themselves):

- In a treatise on celestial navigation by insects, Wehner (1982) describes various experiments on the effect of various test conditions on the celestial compass of dancing bees. With unrestricted vision, the dance directions of the bees relate to the direction of the food source. Obscuring parts of the bees' eyes, or modifying the lighting environment (either a partial view of the natural sky or in polarised light) can result in distributions of dance directions which appear close to uniformity. Other experiments relate to the navigational abilities of ants.

- Batschelet (1981) describes numerous experiments, particularly by K. Schmidt-Koenig, on pigeon homing. In one such experiment by Kiepenheuer (1978), pigeons were kept in three types of cages – cages with walls deflecting the wind clockwise, cages with walls deflecting the wind anti-clockwise, and cages having no deflectors. When subsequently released from a remote location, the vanishing directions of the three groups were significantly different.

Other examples in Batschelet's book and in a more recent book by Upton & Fingleton (1989, Chapter 9) (both of which are veritable celebrations of biological applications) include studies on the orientation of dragonflies with respect to the direction of the sun's rays, the directional behaviour of salamanders in the presence of polarising light, the sun–compass orientation of starhead top-minnows, the orientation of captive robins in the presence of a weak magnetic field, whether sea stars could still move towards the shore after being transported to another habitat at the same depth, and vanishing directions of mass migrations of spiny lobsters.

From a statistical viewpoint, these problems are of a classical nature, requiring display and summary of a single random sample of data (typically with a single modal group); assessing whether the uniform distribution is a reasonable model for the data; comparison of two or more such samples; and occasionally some more complicated modelling of the regression (or generalised linear) model type. By and large, satisfactory approaches now exist for all of them.

The situation is somewhat different for geological data where, for many common problems, adequate methods of analysis are as yet unavailable. A single sample of directions may now manifest several modal groups: indeed determining the number of these groups is often of prime importance. Another important characteristic of much geological data is the presence of a spatial component associated with each directional measurement. This can occur when data are collected across a succession of layers (and so may be amenable to a model incorporating components of variation) or, more commonly, when measurements are made at a variety of spatial locations. An additional layer of complexity is provided when a measurement of orientation has an associated measurement of length. The general issue underlying all of these problems is one of sampling.

Deciding what, and how, to sample is as crucial in Geology as in any other discipline. That it is poorly understood is due in part to the relative lack of attention paid to the Earth Sciences by statisticians, and in part to the difficulties involved in clarifying the constituent issues. Watson (1966) noted:

> Since we are about to give a brief account of formal methods, it is worthwhile reiterating that they will give results that are only as sound as the assumptions they rest on. *Thus the most important part of the analysis precedes the application of these methods.* This initial phase includes such matters as the precise formulation of the problem, decisions about what populations are really to be samples, whether the samples are random, whether the observations are ... directions (i.e. unit vectors), or just axes (i.e. directions without sense ...), thoughts about the relative magnitudes of variability due to measurement errors, site-to-site and exposure-to-exposure variation, etc.

To illustrate these remarks, if not to suggest solutions, we describe a common sampling situation in Structural Geology. (For a much deeper exploration of this subject, see the wide-ranging article by Watson, 1970.)

Consider a geologist seeking to understand the structural fracture pattern in a given region. To start with, data have to be gathered from a number of locations ... but which locations? The required data (in this example) are measurements of fractures in outcropping rocks or in cuttings through rocks. These locations have been determined at the whim of Nature, or of road or railway engineers, and certainly not at the whim of a random sampling mechanism used by a geologist.

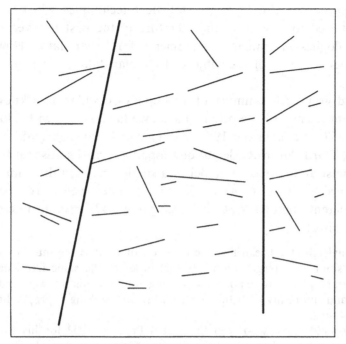

Fig. 1.4 A schematic diagram showing a number of different fracture sets at a sampling site. Typical sampling problems relate to the method of treating major and minor fractures, and to taking account of differing lengths of fracture.

Next, a given exposure may present any or all of the following challenges:

- Some fracture sets not observable because of the dip of the exposed plane
- A very large number of fractures are present: how should a 'representative sample' be chosen? (Some progress can be made with this aspect through statistical methods for sampling.)
- Several distinct (i.e. differently oriented) fracture sets are present, one of them represented by one or two huge (long/wide/deep) fractures, the others by numerous small fractures which can be recorded as several per metre: a schematic view of such an exposure is shown in Figure 1.4. How should relative frequencies be decided? Should allowance be made for fracture length? (Which is the dominant set: indeed, how should 'dominant set' be defined?)
- The 'grain' of the rock is such that it fractures preferentially in certain directions.

The geological/statistical issues are intertwined with DNA-like complexity in this example, and much careful collaborative work is required to achieve a workable approach. The types of statistical techniques which will be needed have started to emerge in the literature in the past few years, and indeed are analogues of similar developments for linear data. However, the historical development of the subject of circular data analysis has been rather different.

The standard graphical summary of a sample of circular data known as the rose diagram seems to have originated, as we have noted, with Florence Nightingale (1858). Later in the 19th century, Lord Rayleigh (1880) (John William Strutt, Third Baron Rayleigh) developed a formal statistical test for uniformity against an alternative model of a single preferred direction. This work went unheeded by statisticians for two decades, before reappearing in a totally different context. Watson (1983, pp. 41–42) gave an interesting account of the history:

> Lord Rayleigh (1880) clarified the question of what is the intensity of the superposition of a large number of vibrations of the same frequency but of arbitrary phase by explicitly assuming that the phases were independently and uniformly distributed on $(0, 2\pi)$ and seeking a proper limiting distribution ...
>
> Sir Ronald Ross, who won the Nobel Prize in 1902 for his work on Malaria, needed to know how the density of mosquitoes would fall as the distance from their breeding place increased ... Though Ross obtained a solution, he asked Karl Pearson for assistance in 1904. Pearson (1905, July 27) asked readers of *Nature* if they could solve the problem of random walk in the plane for *n* steps. Rayleigh (1905, August 3) gave his 1880 asymptotic solution. Kluyver (1906) responded with an elegant study using an integral representation of a "discontinuous factor"... Pearson (1906) numerically evaluated this integral [Kluyver's solution] ... Rayleigh (1919) returned to the topic and treated it in great detail ...

A graph related to Florence Nightingale's rose diagram was introduced into the geological literature by Schmidt (1917) as a way of displaying axial data. Schmidt grouped the data into 5° intervals and then, in each 5° arc, plotted a point at a radial distance equal to the relative frequency of data in that interval; finally, the points were joined up. Examples of Schmidt's graphs are shown in Figure 1.5. However, there was an important difference between Nightingale's diagram and Schmidt's version: in the former (cf. Figure 1.3), sector *area* was plotted proportional to relative frequency, whereas in the latter, sector *radius* was plotted proportional to relative frequency, resulting in a perceptual bias in favour of larger frequencies. Schmidt also compared his diagrams with theoretical plots based on a bipolar or axial version of the Wrapped Normal distribution, as shown in Figure 1.6. A brief account of this aspect of Schmidt's work can be found in Chayes (1949), who also

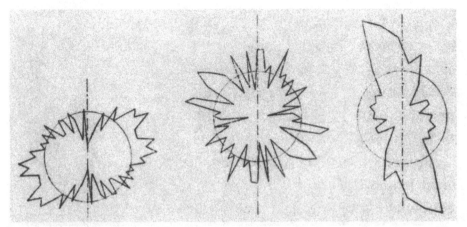

Fig. 1.5 Graphs used by Schmidt (1917) to display samples of axes. The data were grouped into 5° intervals and then, in each 5° arc, a point plotted at a radial distance equal to the relative frequency of data in that interval; finally, the points were joined up. (Bodleian Library, RSL. Gen. Per. 55, p. 528 and opp. p. 538.)

noted that it was extended by Fischer (1926). Today, Schmidt is probably best-remembered for introducing the equal-area projection of spherical data onto the plane (the Schmidt net as it is commonly known).

With one significant exception, early 20th Century experimentation with statistical methods appropriate to circular data took place largely in the geological literature. This exception was a paper by von Mises (1918), who introduced the probability distribution now named after him in connection with modelling experimental errors arising from determinations of supposedly integer atomic weights. Despite the tenuous connection of this application to circular data analysis, the von Mises distribution is now the basis of parametric statistical inference for circular data.

Elsewhere, Reiche (1938) made use of what would now be called CUSUM charts to determine when a sufficient number of measurements have been taken, by plotting ' ... cumulative curves of the vector direction and the consistency ratio [mean resultant length] ... Hence, when the cumulative vector direction curve flattens and remains within a 5° range, it is likely that the sampling is reasonably adequate ...' (Reiche, 1938, p. 913). An example of one of Reiche's charts is shown in Figure 1.7. Reiche's data were actually (three-dimensional) measurements of poles to cross-lamination attitudes, and the data plotted – dip directions – were weighted by the degree of dip.

A year later, Krumbein (1939) introduced the key idea of analysing *axial* data (i.e. data measured as undirected lines) by transforming them to vectors

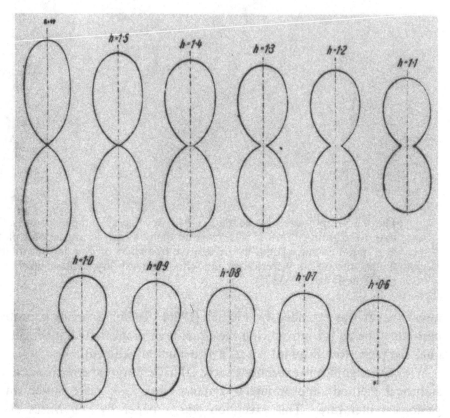

Fig. 1.6 Bipolar Wrapped Normal distributions investigated by Schmidt (1917) as possible models for the data sets in Figure 1.5.

and then back-transforming the results. Krumbein attributed this suggestion to T. L. Page.

Because of the intrinsically spatial character of structural data, geologists experimented with ways of incorporating this feature into data summaries. For example, Potter (1955, Figure 12) displayed smoothed directions on a regular grid superimposed on a map of Western Kentucky, U.S.A. The smoothed direction at each gridpoint was obtained by calculating the mean direction of the data in the four adjacent grid cells.

However, the overwhelming preponderance of these articles, geological or otherwise, were concerned with discussing ways of testing the null hypothesis of uniformity, or the value of using the mean resultant length as a measure of strength of the sample mean direction (see Steinmetz (1962) and Potter & Pettijohn 1977, pages 262–7, for references to much of the geological

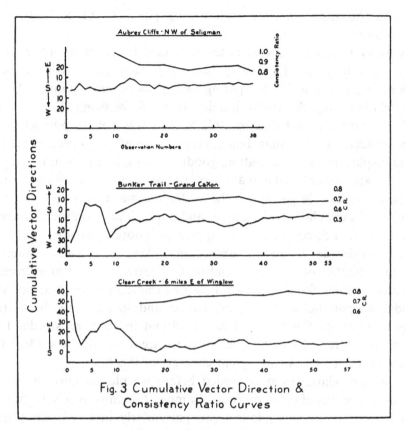

Fig. 3 Cumulative Vector Direction &
Consistency Ratio Curves

Fig. 1.7 CUSUM charts developed by Reiche (1938), to assist in determining when a sufficient number of measurements had been made, provide a stable estimate of the mean direction. (Reproduced by permission of the Syndics of the Cambridge University Library.)

literature). Elsewhere, Salvemini (1940, 1942) developed what is now called circular variance, to study departures from uniformity of monthly deaths in Rome during 1925–7, for various age groups. Interest in tests for uniformity based on runs or spacings has continued to the present day, with numerous contributions from R. J. Beran, J. S. Rao (S. Rao Jammalamadaka) and others.

In the 1950s, statisticians had begun fitting circular probability distributions to data. E. J. Gumbel investigated the suitability of the von Mises distribution for modelling data sets as diverse as monthly egg production, sales of Misses' coats, and monthly occurrences of rainfall exceeding 1 inch per hour (Gumbel 1954). However, significant development of statistical methods for analysing circular data can be reasonably asserted to date from

1956, with the appearance of a pioneering paper by G. S. Watson and E. J. Williams.

This paper introduced methods for statistical inference about the mean direction and dispersion for a single sample of data from a von Mises distribution, and methods for comparing two or more such samples. Further developments during the 1960s, notably by G. S. Watson, M. A. Stephens and K. V. Mardia, meant that by 1972 Mardia was able to publish a comprehensive account of circular data analysis systematically covering methods for data display and summarisation, goodness of fit, and nonparametric and parametric approaches to the analysis of a single sample and of two or more samples. (Curiously, large-sample theory for the mean direction, leading to the notion of a circular standard error and approximate confidence intervals for the true mean direction, did not appear for another decade.)

In the two decades following publication of Mardia's book, advances occurred in a number of areas: correlation and regression and more recently, time series analysis for circular data; large-sample methods and bootstrap methods; nonparametric density estimation and spatial smoothing. (References to much of this material can be found in Jupp & Mardia 1989.) However, much remains to be done in these areas, as will be evident from the material presented in subsequent chapters of this book.

One word of clarification is needed before we plunge into our subject. It relates to the meaning of the word 'sample' (or more precisely, 'random sample'). This is an ambiguous term, particularly in the Earth Sciences where it is sometimes taken to mean a single measurement. Throughout this book, it will always be taken to mean a set of (statistically) independent measurements of a single phenomenon, such as the vanishing directions of homing pigeons each treated and released under the same experimental conditions, or a set of randomly selected measurements of fractures caused by a specific directional fracturing event. Amplification of words and phrases here, and of other statistical terminology and concepts in this book, can be found in the glossary of statistical terms provided in Fisher, Lewis and Embleton (1987, §2.3) or more generally in Marriott (1990).

2

Descriptive methods

2.1 Introduction

In the Preface, the various types of circular data – vectors/axes, uniform/unimodal/multimodal – were described together with the usual ways of recording them. This chapter is concerned with the most basic aspects of statistical analysis of a single sample of circular measurements $\theta_1, ..., \theta_n$: methods for displaying the sample, and simple summary quantities which can be calculated from the sample.

2.2 Data display

Why do we display data? There is no shortage of reasons, but among the most important are:

- to gain an initial idea of the important characteristics of the sample; for example, does the sample appear to be from a uniform (or isotropic) distribution; or from a unimodal distribution; or from a multimodal distribution?
- to emphasise such characteristics
- to suggest models for the data, such as a von Mises model for a sample which appears to be drawn from a symmetric unimodal distribution
- to try to avoid doing something stupid: note the advice once given to students graduating from a weather-forecasting course – before issuing a forecast, look out of the window.

In view of the first of these points, the nature of any display, apart from a simple plot of the raw data, will depend on the number of modal groups apparent in the sample. Unless the sample is clearly unimodal, constructing a useful display is not a trivial exercise. Some forms of display are described below, others will be encountered in Chapters 5, 6 and 7.

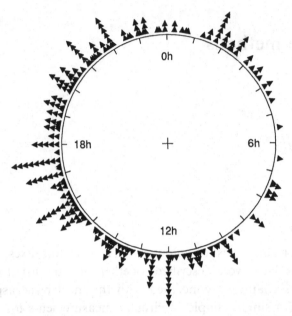

Fig. 2.1 Arrival times on a 24-hour clock of 254 patients at an intensive care unit, over a period of about 12 months. See Example 2.1.

(i) Raw data plot

Raw data plots are important, particularly because of the last reason given above for displaying data. Unfortunately, they are frequently omitted. Other data displays of the histogram variety can conceal small but important features of the data distribution such as the existence of a small modal group (that is, an isolated concentration of points), or an outlying point whose existence may be of specific interest. For axial data, it is generally helpful either to plot a point at both ends of each axis in the sample, or to plot the axes as diameters, otherwise features of the data may be lost because of an arbitrary choice of 180° arc in which to display them (arbitrary, that is, with respect to the location of the main data mass).

Example 2.1 Figure 2.1 shows a plot of arrival times on a 24-hour clock of 254 patients at an intensive care unit, over a period of about 12 months. (The data are listed in Appendix B1.) The arrival times may be regarded as circular measurements, with an arrival time of m minutes after midnight corresponding to a circular measurement of $360 \times m/(24 \times 60)$ degrees; thus 1° corresponds to 4 minutes of time. The plot suggests a fairly steady stream of arrivals between 10am and 10pm, with smaller clusters of arrivals about 2am and 7am.

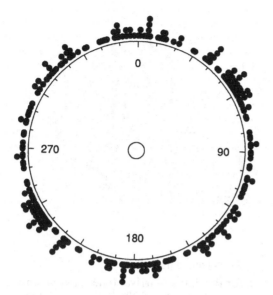

Fig. 2.2 Measurements of long-axis orientations of 133 feldspar laths in basalt. See Example 2.2.

Example 2.2 Figure 2.2 shows a plot of measurements of long-axis orientation of 133 feldspar laths in basalt. Both ends of each axis are plotted. There appear to be a number of modal groups present. (The data are listed in Appendix B2.)

(ii) Histograms

There are two basic forms: linear and angular.

Linear histograms

A linear histogram is constructed as if the data were linear, not angular. We select some starting point (e.g. 0° if the data are recorded in the range (0°, 360°), or −180° if they in the range (−180°, 180°)) and some binwidth or grouping interval (say 5°, 10° or 20°) and create a histogram in the usual way.

Example 2.3 A histogram for the data in Figure 2.1 is shown in Figure 2.3, using a binwidth of 20° (80 minutes). The features noted in Example 2.1 are again evident.

Angular histograms I

A simple angular histogram is obtained by wrapping a linear histogram around a circle. Each bar in the histogram is centred at the midpoint of

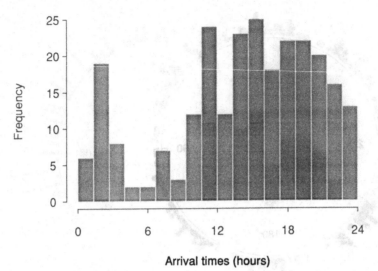

Fig. 2.3 A linear histogram for the data on arrival times at an intensive care unit shown in Figure 2.1, using a binwidth of 20° (i.e. 1 hour 20 minutes).

its grouping interval, with the length of the bar proportional to the relative frequency (or the amount of data) in the group.

Example 2.4 The data of Figure 2.1 are plotted as an angular histogram in Figure 2.4, using a binwidth of 20° (80 minutes).

Angular histograms II

A more commonly used method is the *rose diagram*, in which each group is displayed as a sector. The radius of each sector is probably best taken to be proportional to *the square root of* the relative frequency of the group, so that the *area* of the sector is proportional to the group frequency.

Example 2.5 Figure 2.5(a) shows a rose diagram for the data of Figure 2.1, with the radius of each sector proportional to the square root of the relative frequency of the groups, and using an angular binwidth of 20° (80 minutes). If the rose diagram is plotted with radius proportional to frequency (and the same binwidth), we get the picture in Figure 2.5(b), in which the group of arrivals around 2am is considerably de-emphasised and the isolated arrivals between 5am and 7am would pass unnoticed without the aid of the accompanying plot of raw data.

(Potter & Pettijohn (1977 p. 375), note that in the context of palaeocurrent analysis, the rose '... conventionally indicates the direction *toward which current moved* unlike the wind rose of the meteorologists, which indicates the direction from which the winds came'.)

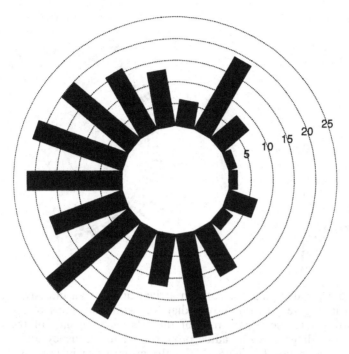

Fig. 2.4 Angular histogram for the data on arrival times at an intensive care unit shown in Figure 2.1, using a binwidth of 20° (i.e. 1 hour 20 minutes). See Example 2.4.

There are several points to consider in relation to histogram displays.

- Histograms are easy to construct. In particular, almost any statistical package can produce a reasonable linear histogram.
- Linear histograms involve an arbitrary choice of starting point. This defect can be remedied by repeating a complete cycle of the histogram.

> **Example 2.6** A version of Figure 2.3, modified to allow the eye to assess the circular distribution better, is shown in Figure 2.6, by repeating the complete 24-hour cycle.

- More importantly, use of histograms involves an arbitrary choice of cell or group boundaries. An unfortunate choice of boundary can result in serious distortion of the information in the sample about the number, sizes and locations of modal groups.

> **Example 2.7** Figure 2.7 shows rose diagrams for the data of Figure 2.2, with the cell boundaries in Figure 2.7(b) calculated starting at 15° instead of 0° as used in Figure 2.7(a). There is a noticeable difference in shape

(a)   (b)

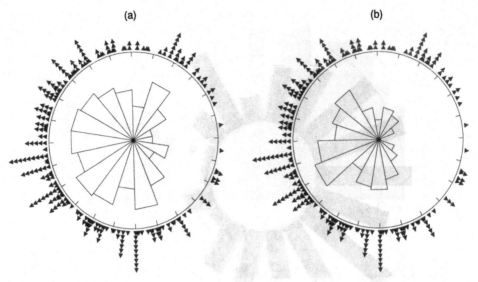

Fig. 2.5 Rose diagrams for data on arrival times at an intensive care unit shown in Figure 2.1. (a) A rose diagram with the radius of each sector proportional to the square root of the relative frequency of the groups. (b) A rose diagram with radius proportional to frequency and the same binwidth as in (a). In Figure 2.5(b), the group of arrivals around 2am is considerably de-emphasised and the isolated arrivals between 5am and 7am would pass unnoticed without the aid of the accompanying plot of raw data.

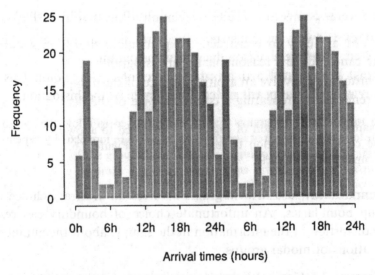

Fig. 2.6 A version of Figure 2.3, modified to allow the eye to assess the circular distribution better, by replication of a complete cycle of the data.

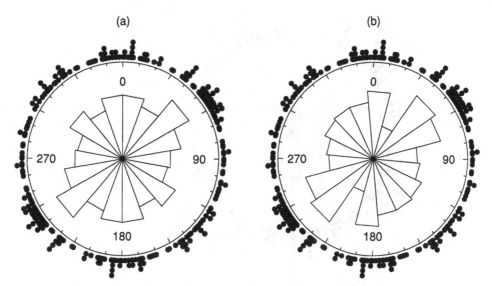

Fig. 2.7 Rose diagrams of the long-axis orientations of feldspar axes displayed in Figure 2.2, showing the effect of differing locations of bin boundaries. (a) Bin boundaries start at 0°. (b) Bin boundaries start at 15°. See Example 2.7.

between the two diagrams, which were both derived from the same data set with the same amount of smoothing (that is, same binwidth). (For an even more striking illustration of this effect for linear data, see Fisher 1989.)

- The other choice to be made is of the binwidth of each cell. In view of the preceding comments about the poor performance of rose diagrams in revealing the shape of the underlying distribution, we simply suggest using an arbitrary size of 10° or 20° (or possibly more for very small data sets) if, indeed, a display of this sort is needed.

Example 2.8 Rose diagrams corresponding to Figures 2.5 and 2.7(a) using a binwidth of 10° rather than 20° are shown in Figures 2.8 and 2.9 respectively. The increased 'spikyness' of the plots makes it more difficult to grasp the broad distributional features, with little apparent gain in exchange for this.

The key point to be made is that the principal virtue of histogram-type displays *is* their simplicity of construction. In other ways, they pose the same question as do the smooth nonparametric density estimates recommended in *(iv)* below – namely, how much smoothing is required, expressed as choice of binwidth – and in addition they have the unsavoury ability to manifest quite different shapes depending on the location of the cell boundaries. A

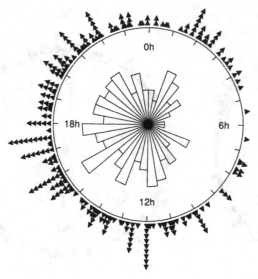

Fig. 2.8 The effect of using a smaller binwidth. The data on arrival times plotted as a rose diagram with binwidth 20° in Figure 2.5 are plotted here as a rose diagram with binwidth 10°, giving a plot with many more modes. See Example 2.8.

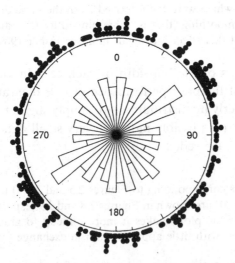

Fig. 2.9 Rose diagram of the long-axis orientations of feldspar axes displayed in Figure 2.2, using a binwidth of 10°, compared with the use of a 20° binwidth in Figure 2.7(a). The result of decreasing the binwidth is an increasingly spiky plot, with no apparent gain in terms of appreciation of broad distributional features. See Example 2.8.

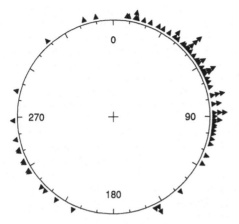

Fig. 2.10 Directions taken by 76 turtles, after treatment. See Example 2.9.

far more powerful and reliable form of display for enhancing a plot of the raw data is furnished by a nonparametric estimate of density such as that described in *(iv)*.

(iii) Stem-and-leaf diagrams

These semi-graphic displays are essentially histograms retaining the raw data values, and their mode of construction is immediately evident from an example.

> **Example 2.9** Figure 2.10 comprises measurements of the directions taken by 76 turtles after treatment. (The data are listed in Appendix B3.) From the discussion in Stephens (1969b, p. 20) '... it is thought that the turtles have a preferred direction but some are confusing forwards with backwards'; the raw data plot clearly suggests some degree of bimodality. Linear and angular stem-and-leaf diagrams for these data are shown in Figures 2.11 and 2.12.

As can be seen from this example, the construction is simple: each stem corresponds to the data falling in, say, a 10° interval, with the leaves being the data values in that interval, sorted into increasing order and discarding 10's and 100's digits. As with linear histograms, the linear stem-and-leaf diagram can be extended in each direction if necessary (cf. Figure 2.6).

A stem-and-leaf diagram is a very convenient device for displaying data without losing the individual measurements. Note however, that if data are actually gathered and *recorded* using a stem-and-leaf diagram, some possibly important information is lost, namely the time-order in which the data were gathered.

```
 0 :  8,  9
 1 :  3,  3,  4,  8
 2 :  2,  7
 3 :  0,  4,  8,  8
 4 :  0,  4,  5,  7,  8,  8,  8, 8
 5 :  0,  3,  6,  7,  8,  8
 6 :  1,  3,  4,  4,  4,  5,  5, 8
 7 :  0,  3,  8,  8,  8
 8 :  3,  3,  8,  8,  8
 9 :  0,  2,  2,  3,  5,  6,  8
10 :  0,  3,  6
11 :  3,  8
12 :
13 :  8
14 :
15 :  3,  3,  5
16 :
17 :
18 :
19 :
20 :  4
21 :  5
22 :  3,  6
23 :  7,  8
24 :  3,  4
25 :  0,  1,  7
26 :  8
27 :
28 :  5
29 :
30 :
31 :  9
32 :
33 :
34 :  3
35 :  0
```

Fig. 2.11 Linear stem-and-leaf diagram for the directions taken by 76 turtles in Figure 2.10.

(iv) Nonparametric density estimates

A wide range of methods fall under this heading as can be seen, for example, by consulting Silverman (1986). Here, we confine ourselves to a simple form known as a *kernel density estimate*, which can be thought of as a sort of moving average. In essence, we smear out the contribution $(1/n)$ of each data point over a (relatively small) arc containing that data point; the density estimate in a given direction θ is then the sum of the contributions of the smeared-out points at θ. Figure 2.13(a) is a raw data display of six measurements, with the small bumps representing the smearing-out of the 'weight' $(1/6)$ of each point; the overall density estimate is just the sum of all these individual contributions, and is shown in Figure 2.13(b).

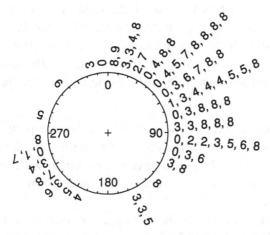

Fig. 2.12 Angular stem-and-leaf diagram for the directions taken by 76 turtles in Figure 2.10.

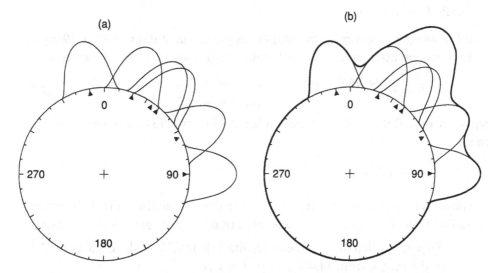

Fig. 2.13 Nonparametric density estimation using the kernel method. (a) Each data point is smeared out using a small distribution. (b) The nonparametric density estimate is the sum of these small contributions.

Mathematically, the density estimate is written as

$$\hat{f}(\theta) = (nh)^{-1} \sum_{i=1}^{n} w\left(\frac{\theta - \theta_i}{h}\right) \qquad (2.1)$$

where w is a function like one of the bumps in Figure 2.13(a) and h controls the amount of smearing-out, or *smoothing*, that is to be done.

The key factor determining the behaviour of the density estimate is not the precise shape of the function w but the magnitude of h: the larger h is, the greater the amount of smoothing and consequently the fewer the number of modes or 'bumps' in $\hat{f}(\theta)$. The amount of smoothing (i.e. the size of h) has to be related to the sample size n and to the dispersion in the data, for the following reasons:

- dependence on n – the more data we have, the more we are prepared to search for details of the true underlying density $f(\theta)$; with a small sample, we have little information about $f(\theta)$, and have to be content with drawing simple inferences, for example, that the sample has probably been drawn from a unimodal distribution.
- dependence on dispersion – clearly, if the data are confined to a 30° arc, we are operating on a different scale from when the data are spread over 180°. (A more subtle feature emerges if the data are clumped in several groups in the 180° arc. A special approach is recommended below to handle this situation.)

The following procedure is an adaptation to circular data (Fisher 1989) of linear data methods in Silverman (1986), using a quartic kernel function

$$w(\theta) = \begin{cases} 0.9375(1 - \theta^2)^2 & -1 \le \theta \le 1 \\ 0 & \text{otherwise} \end{cases} \tag{2.2}$$

$w(\theta)$ is a probability density function, so the factor 0.9375 appears to ensure that

$$\int_{-1}^{1} w(\theta)\, d\theta = 1$$

Suppose first that the data are in a single clump; modifications for multi-modal or symmetric unimodal data sets are described later in this section.

Step 1. Calculate the mean resultant length $\bar{R}$ of the data as in **§2.3**, and hence $\hat{\kappa}$ from (4.40) and (4.41). Let

$$\hat{\zeta} = 1/\hat{\kappa}^{\frac{1}{2}} \tag{2.3}$$

Step 2. Calculate

$$h_0 = 7^{\frac{1}{2}}\hat{\zeta}/n^{\frac{1}{5}} \tag{2.4}$$

(For a single group of data, this parameter is designed for use with the quartic kernel function in (2.2).)

Step 3. For any given direction θ and smoothing parameter h, calculate the density estimate $f(\theta)$ using the following algorithm:

3.1 $i = 0$
sum $= 0$

3.2 $i = i + 1$
$d_i = |\theta - \theta_i|$
$e_i = \min(d_i, 2\pi - d_i)$

3.3 If $e_i \geq h$ go to **3.2**

3.4 sum $=$ sum $+ (1 - e_i^2/h^2)^2$

3.5 If $i < n$ go to **3.2**

3.6 $f(\theta) = 0.9375 \times \text{sum}/(n \times h)$

3.7 Repeat this for, say, 100 values of θ equally spaced between 0 and 2π, yielding

$$(\theta_1^*, f_1), \ldots, (\theta_{100}^*, f_{100})$$

where $f_j = f(\theta_j^*)$, $j = 1, \ldots, 100$.

3.8 To plot the density around a circle of radius r, set

$$f_j^* = r(1 + \pi f_j)^{\frac{1}{2}} - r, \ j = 1, \ldots, 100$$

(where f_j^* is to be measured out from the circle). Join up the points

$$(\theta_1^*, f_1^*), \ldots, (\theta_{100}^*, f_{100}^*), (\theta_1^*, f_1^*) \tag{2.5}$$

to give a picture of the smoothed density. f^* is used instead of f to reduce distortion in the plot, as with rose diagrams. This improves on the formula in Fisher (1989).

Step 4. It is usually helpful to look at the density estimates corresponding to values of h in the range $(0.25h_0, 1.5h_0)$.

More efficient algorithms can be based on the fast Fourier transform (see e.g. Silverman 1986) or by using binning (see e.g. Scott & Sheather 1985).

Modifications for multimodal data. More care is needed when the data appear to have two or more modal groups or clumps. Whilst it is difficult to formulate a simple general procedure, satisfactory results can often be obtained by calculating $\bar{R}$ in Step 1 just from the data in the largest clump, and using as the value of n in Step 2 the number of data points in that clump. (Note: Fisher (1989) did not make explicit the need to reduce the value of n to that of the largest clump.) The other point to note is that displaying the density estimate in a linear form (suitably extended, as in Example 2.6 for histograms) may well aid perception of the number of modal groups. The value of the circular density estimate lies partly in relating these groups to each other and to the data, in a compact form. This issue will be explored in the next example.

When the data are axial, or undirected, such as those in Figure 2.2, the following adjustments are required:

Step 0.   Convert the data to *vectors*, or directed lines, by doubling them.

Replace *Step 3.8* by

Step 3.8*.   Join up the points

$$(\tfrac{1}{2}\theta_1^*, f_1^*), \ldots, (\tfrac{1}{2}\theta_{100}^*, f_{100}^*), (\tfrac{1}{2}\theta_1^*, +\pi, f_1^*)$$
$$, \ldots, (\tfrac{1}{2}\theta_{100}^*, +\pi, f_{100}^*), (\tfrac{1}{2}\theta_1^*, f_1^*) \tag{2.6}$$

Example 2.10   To construct a nonparametric density estimate for the data of Figure 2.1, suppose we assume initially that the sample has been drawn from a unimodal distribution. Following the algorithm described above, we obtain $\bar{R} = 0.3226$, $\hat{\kappa} = 0.68$, and finally $h_0 = 1.06$. The density estimate corresponding to this amount of smoothing is shown in Figure 2.14(a), together with a linear version in Figure 2.14(b). Too much of the structure of the data appears to have been smoothed out. Accordingly, we can look at, say, the picture corresponding to $0.5 \times h_0 = 0.53$. This is shown in Figures 2.14(c) and 2.14(d). Now, a small peak near 2am is evident, as are small peaks about 11.30am and 6.00pm, but a possible mode near 7am which was noted in Example 2.1, is still hard to see. If we reduce the amount of smoothing even further – using $0.25 \times h_0 = 0.26$ – we get the pictures in Figure 2.14(e) and 2.14(f). The possible small mode at 7am is now present, but so are a few other bumps atop the main modal group. At this point, it is a question of interpretation. Is there a reason why there should be several local increases in the density (they do not appear to have a fixed time period between them), or are they simply artefacts arising from undersmoothing the data? In the absence of other information or a formal statistical test for the number of modes, it is difficult to proceed further. We conclude this analysis by looking at what happens if the modification for multimodal data sets is followed. Suppose we decide that there is a main group between 10am and 10pm, and compute $\bar{R}$ just from the 188 values in that group. We get $\bar{R} = 0.6358$, leading to $h_0 = 0.72$, which gives effectively the same display as Figure 2.14(b). All things considered, it seems reasonable to present Figure 2.14(c), *with the raw data plot of Figure 2.1 included*, as a summary of the qualitative features of the data.

Modifications for symmetric data. If there is reason to believe that the sample has been drawn from a unimodal distribution which is symmetric about its mean (equivalently, median) direction, this information can be

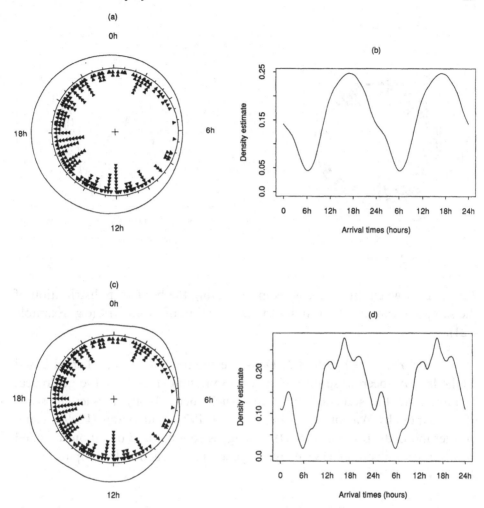

Fig. 2.14 Effect of choice of smoothing parameter *h* on a nonparametric density estimate for the arrival times in Figure 2.1. (a) Oversmoothing (*h* too large) can lead to loss of important structure. In this plot, $h = 1.06$. (b) a linear version (a), extended at each end to allow the eye to assess the circular distribution better. (c) Reducing the amount of smoothing allows more subtle features to emerge, such as the small mode near 2.00am, and the minor modes each side of the principal mode. In this plot, $h = 0.53$. (d) a linear version of (c). (e) Undersmoothing (h too small) can lead to too many modes. In this plot, $h = 0.26$. (f) a linear version of (e).

used in calculating the nonparametric density estimate. Calculate the sample mean direction $\bar{\theta}$ as in (2.9) below. Then calculate the density estimate based on $2n$ data points

$$\theta_1, \ldots, \theta_n, 2\bar{\theta} - \theta_1, \ldots, 2\bar{\theta} - \theta_n$$

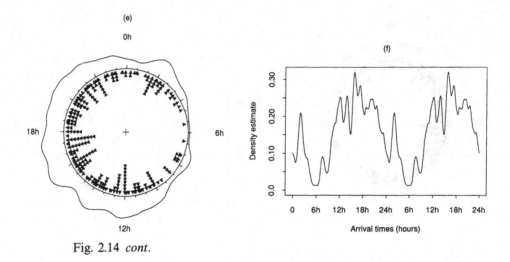

Fig. 2.14 *cont.*

An example when this can arise is in displaying the bootstrap distribution of the sample mean direction under the assumption of symmetry (e.g. Example 4.21).

References and footnotes. Density estimates, particularly kernel-based methods, have been in sporadic use for a long time, mostly in the geological literature. In the statistical literature, such methods are implicit in some of the papers by Watson (e.g. Watson 1969, 1971) and Beran (1969a) using Fourier methods. Boneva *et al.* (1971) suggested a method based on so-called 'histosplines'. Silverman (1986) gives a general introduction to nonparametric density estimation.

2.3 Simple summary quantities

In the following chapter, a number of basic quantities (or *parameters*) are introduced which describe important features of the population from which data have been sampled, such as the reference directions and dispersions of modal groups. Here, we describe ways of summarising sample information about these population parameters.

2.3.1 Sample trigonometric moments

A basic concept is that of a *sample moment*. Suppose that our sample appears to be drawn from a distribution with a single modal group, such as that in Figure 4.11. Can we estimate the reference direction of the mode by the

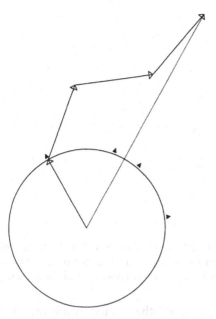

Fig. 2.15 Example of addition of unit vectors: the resultant vector has the direction of the mean direction of the individual vectors, and its length is their resultant length. The individual circular values corresponding to the unit vectors are also shown.

usual arithmetic mean $\sum_{i=1}^{n} \theta_i / n$? Clearly, the answer is no – e.g. consider the sample of three points 359°, 1°, 3°. (This is the notorious cross-over problem familiar to meteorologists.) In fact, the natural way to combine (unit) vectors is to use vector addition, as shown in Figure 2.15 for a simple case of four data points; note that the order of the sequence of vectors is immaterial. Calculate

$$C = \sum_{i=1}^{n} \cos \theta_i, \quad S = \sum_{i=1}^{n} \sin \theta_i, \quad R^2 = C^2 + S^2 \quad (R \geq 0). \qquad (2.7)$$

The direction $\bar{\theta}$ of the vector resultant of $\theta_1, \ldots, \theta_n$ is given by

$$\cos \bar{\theta} = C/R, \qquad \sin \bar{\theta} = S/R \qquad (2.8)$$

or

$$\bar{\theta} = \begin{cases} \tan^{-1}(S/C) & S > 0, C > 0 \\ \tan^{-1}(S/C) + \pi & C < 0 \\ \tan^{-1}(S/C) + 2\pi & S < 0, C > 0 \end{cases} \qquad (2.9)$$

and is known as the *mean direction*.

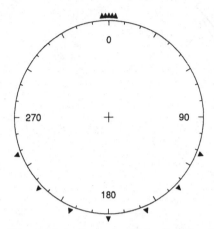

Fig. 2.16 A data set with well-defined modal groups yet with mean resultant length effectively zero. So the mean resultant length is not necessarily a good guide to the concentration or lack of concentration of a circular sample.

The quantity R is the *resultant length* of the vector resultant, and lies in the range $(0, n)$. The *mean resultant length* $\bar{R}$ associated with the mean direction $\bar{\theta}$ is defined by

$$\bar{R} = R/n \tag{2.10}$$

and lies in the range $(0, 1)$. Its extreme values have some interesting properties. $\bar{R} = 1$ implies that all the data points are coincident. However, $\bar{R} = 0$ does *not* imply uniform dispersion around the circle: for example, the data in Figure 2.16 have $\bar{R} = 0$, yet the data distribution is clearly highly structured. So $\bar{R}$ is not necessarily a useful indicator of dispersion or spread of the data unless they comprise a single group.

The quantity

$$V = 1 - \bar{R} \tag{2.11}$$

is defined as the *sample circular variance*. Similarly to the variance of linear data, the smaller the value of the circular variance, the more concentrated the distribution. However, note that $0 \le V \le 1$, unlike an ordinary linear variance. Also, note that in the light of the above remark concerning the interpretation of $\bar{R} = 0$, $V = 1$ does not necessarily imply a maximally dispersed distribution.

The sample *circular standard deviation* is defined as

$$v = \{-2\log(1 - V)\}^{\frac{1}{2}} \tag{2.12}$$

The reasons for defining the circular standard deviation in this way, rather than as the square root of the sample circular variance by analogy with the linear standard deviation, are elaborated in §3, where the corresponding population parameters are discussed. A good approximation to (2.12) is given by

$$v \simeq \begin{cases} (2V)^{\frac{1}{2}} & \text{for } V \text{ small} \\ [2(1-\bar{R})]^{\frac{1}{2}} & \text{for } \bar{R} \text{ large}, \end{cases} \tag{2.13}$$

the error in this approximation being less than 5% for $V < 0.18$, that is, $\bar{R} > 0.82$.

The quantities $\bar{\theta}$ and $\bar{R}$ are the angular and amplitude components of the first (uncentred) *trigonometric moment*

$$m_1' = \bar{C} + i\bar{S} = \bar{R}e^{i\bar{\theta}} \qquad (i = \sqrt{-1}) \tag{2.14}$$

where

$$\bar{C} = C/n, \quad \bar{S} = S/n, \quad \bar{R} = (\bar{C}^2 + \bar{S}^2)^{\frac{1}{2}} \tag{2.15}$$

In terms of the population parameters μ (mean direction) and ρ (mean resultant length) of §3, we can also write

$$m_1' = \hat{\rho}e^{i\hat{\mu}} \tag{2.16}$$

where '$\widehat{}$' denotes a sample estimate for the population quantity it surmounts; and so

$$\hat{\mu} = \bar{\theta}, \quad \hat{\rho} = \bar{R} \tag{2.17}$$

Another important sample trigonometric moment is

$$m_2' = \bar{C}_2 + i\bar{S}_2 = \hat{\rho}_2 e^{i\hat{\mu}_2} \tag{2.18}$$

where

$$\bar{C}_2 = (1/n)\sum_{i=1}^{n}\cos 2\theta_i, \quad \bar{S}_2 = (1/n)\sum_{i=1}^{n}\sin 2\theta_i, \quad \hat{\rho}_2 = (\bar{C}_2^2 + \bar{S}_2^2)^{\frac{1}{2}} \tag{2.19}$$

and in general, we can define the *p*th moment by

$$m_p' = \bar{C}_p + i\bar{S}_p = \hat{\rho}_p e^{i\hat{\mu}_p} \tag{2.20}$$

where

$$\bar{C}_p = (1/n)\sum_{i=1}^{n}\cos p\theta_i, \quad \bar{S}_p = (1/n)\sum_{i=1}^{n}\sin p\theta_i \tag{2.21}$$

and $\hat{\rho}_p$ and $\hat{\mu}'_p$ are calculated from $\bar{C}_p$ and $\bar{S}_p$ using

$$\hat{\rho}_p^2 = \bar{C}_p^2 + \bar{S}_p^2 \tag{2.22}$$

and

$$\hat{\mu}'_p = \begin{cases} \tan^{-1}(\bar{S}_p/\bar{C}_p) & \bar{S}_p > 0, \bar{C}_p > 0 \\ \tan^{-1}(\bar{S}_p/\bar{C}_p) + \pi & \bar{C}_p < 0 \\ \tan^{-1}(\bar{S}_p/\bar{C}_p) + 2\pi & \bar{S}_p < 0, \bar{C}_p > 0 \end{cases} \tag{2.23}$$

The *centred* sample trigonometric moments are obtained by calculating the various moments relative to the sample mean direction $\bar{\theta}$:

$$m_p = (1/n) \sum_{i=1}^{n} \cos p(\theta_i - \bar{\theta}) + \mathrm{i}\,(1/n) \sum_{i=1}^{n} \sin p(\theta_i - \bar{\theta}) \tag{2.24}$$

$$= \hat{\rho}_p \mathrm{e}^{\mathrm{i}(\mu'_p - p\bar{\theta})} \tag{2.25}$$

In particular,

$$m_1 = \hat{\rho} = (1/n) \sum_{i=1}^{n} \cos (\theta_i - \bar{\theta}) = \bar{R} \tag{2.26}$$

(because $\sum_{i=1}^{n} \sin(\theta_i - \bar{\theta}) = 0$), and

$$m_2 = \hat{\rho}_2 = (1/n) \sum_{i=1}^{n} \cos 2(\theta_i - \bar{\theta}) \tag{2.27}$$

From the first and second central trigonometric moments we form another measure of spread, the *sample circular dispersion* $\hat{\delta}$, defined by

$$\hat{\delta} = (1 - \hat{\rho}_2)/(2\bar{R}^2), \tag{2.28}$$

and also measures of *skewness* and *kurtosis* (or peakedness), defined respectively by

$$\hat{s} = [\hat{\rho}_2 \sin (\hat{\mu}_2 - 2\hat{\mu})]/(1 - \bar{R})^{\frac{3}{2}} \tag{2.29}$$

and

$$\hat{k} = [\hat{\rho}_2 \cos (\hat{\mu}'_2 - 2\hat{\mu}) - \bar{R}^4]/(1 - \bar{R})^2 \tag{2.30}$$

The circular dispersion $\hat{\delta}$ plays an important rôle in calculating a confidence interval for a mean direction, and in comparing and combining several sample mean directions. Skewness $\hat{s}$ and kurtosis $\hat{k}$ have rather more descriptive uses. Data from a unimodal symmetric distribution such as the von Mises or Wrapped Normal (see Chapter 3) will tend to have sample kurtosis values around zero; more peaked distributions will tend to have positive sample kurtosis.

The need can arise for an adjustment for the effect of grouping, if the grouping interval is large. Data are commonly measured to the nearest 1°, or 5°, in which case no adjustment is required to the sample mean direction $\bar{\theta}$, or higher trigonometric means $\hat{\mu}_p$. Indeed, the error in not adjusting $\bar{R}$ or $\hat{\rho}_2$ is of the order of 1% or less for grouping intervals as large as 30° for $\bar{R}$ and 15° for $\hat{\rho}_2$. For example, if the measurements being made are times during the day, measured to the nearest hour (equivalently, 15°), it is unlikely to be worthwhile making the adjustment.

One way in which larger grouping intervals can arise occurs with axial data, in which case the grouping interval will double when the data values are converted to vectors for purposes of analysis (cf. **§2.4**). For larger grouping intervals h, use $A(h) \times \bar{R}$ and $A(2h) \times \hat{\rho}_2$ instead of $\bar{R}$ and $\hat{\rho}_2$ respectively, where

$$A(h) = h/(2 \sin (h/2)) \tag{2.31}$$

and $\hat{\rho}_2$ is defined by (2.19) (see e.g. Mardia 1972a, §2.8). The definitions of skewness and kurtosis are due to Mardia (1972a, §2.7.2).

2.3.2 Other measures of location and spread

Apart from the sample mean direction, there are two other measures of location or reference direction sometimes used in applications. These are the *sample median direction* and the *sample modal direction*.

A simple way to explain the sample median $\tilde{\theta}$ is by using a picture. (The definition is given in (2.32).) Figure 2.17 shows separately the cases of odd and numbers of data points (Figures 2.17(a) and 2.17(b) respectively). We choose an axis which divides the data into two equal groups (a *median axis*) and then select the end giving the smaller value of $\tilde{\theta}$ in (2.32). When the sample size is odd, the median direction passes through a data point; when it is even, we take the midpoint between two points. Note that with multimodal or isotropic data, the median direction may not be uniquely defined; however it is unique for reasonably unimodal data.

The value of $\tilde{\theta}$ can be found as follows. If the data are concentrated on an arc of the circle substantially less than the complete circle, an algorithm for linear data can be used. If the data mass overlaps 0°, make an initial rotation of the data so that this is not the case, apply the algorithm, then rotate the answer back. Otherwise, $\tilde{\theta}$ can be found by minimising the function

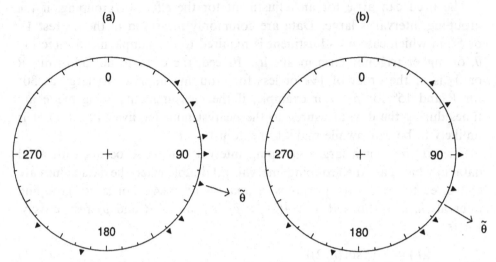

Fig. 2.17 Definition of the sample median direction. (a) Sample size odd: middle point selected. (b) Sample size even: centre of arc between two middle points selected.

$$d(\theta) = \pi - \frac{1}{n} \sum_1^n |\pi - |\theta_i - \theta|| \tag{2.32}$$

The *sample modal direction* $\check{\theta}$ is the direction corresponding to the maximum concentration of the data. (More generally, *any* direction corresponding to a local maximum concentration of the data is a modal direction.) Unfortunately, this is not a well-defined direction for a sample of discrete measurements. In the past, $\check{\theta}$ has been determined from the rose diagram as the midpoint of the cell with largest frequency. However, as we have seen in §**2.2**(*ii*), particularly in Example 2.7, the modal group can vary considerably in a rose diagram, depending on the locations of cell boundaries and on the amount of smoothing (cell width). One way of determining $\check{\theta}$ would be as the value of θ maximising a density estimate $\hat{f}(\theta)$, although this method also depends on the amount of smoothing. Of course, such a vector can be determined for each modal group.

Turning to measures of spread, we define the *sample range* as the shortest angular measure of arc encompassing all of the data. A measure of spread associated with the median direction $\tilde{\theta}$ is the *mean deviation*

$$\pi - (1/n) \sum_{i=1}^n |\pi - |\theta_i - \tilde{\theta}||, \tag{2.33}$$

that is, the minimum value obtained by $d(\theta)$ in (2.32).

2.3.3 *Linear order statistics and circular ranks*

If we treat the sample $\theta_1, \ldots, \theta_n$ as *linear* data and re-arrange them into ascending order to get

$$\theta_{(1)} \leq \cdots \leq \theta_{(n)} \tag{2.34}$$

say, then $\theta_{(i)}$ is termed the ith *order statistic*. Clearly, this ordering changes in cyclic fashion as the choice of zero direction changes. Nevertheless the linear order statistics will be of some value in later chapters.

Again regarding the data as linear, let r_i be the *rank* of θ_i among $\theta_1, \ldots, \theta_n$ (rank 1 being the smallest) : that is, θ_i corresponds to the order statistic $\theta_{(r_i)}$. The *circular rank* of θ_i is then defined as

$$\gamma_i = 2\pi r_i/n \;\;, \quad i = 1, \ldots, n \tag{2.35}$$

γ_i is also known as the *uniform score* corresponding to θ_i.

The circular ranks depend on the choice of zero direction but, like the linear order statistics, will also be of value in later chapters.

2.4 Modifications for axial data

As this book is concerned only with *statistical analysis* of data, rather than with *statistical theory*, the recommendations suggested below are rather limited in their scope. The simplest approach is to transform the axial or *p*-axial data to vector data –

$$\theta_i \longrightarrow \begin{cases} 2 \times \theta_i \; [\text{modulo } 360°] & \text{for axial data} \\[2ex] p \times \theta_i \; [\text{modulo } 360°] & \text{for } p\text{-axial data} \end{cases} \tag{2.36}$$

– analyse the vector data as required, and back-transform the results. Back-transformation applies mainly to finding the mean or median axis. It is recommended that measures of spread be left in vectorial units.

An alternative definition for a reference direction which has been recommended for axial data is the principal axis (eigenvector associated with the largest eigenvalue) of the matrix

$$\begin{bmatrix} \sum_{i=1}^n \cos^2 \theta_i & \sum_{i=1}^n \sin \theta_i \cos \theta_i \\ \sum_{i=1}^n \sin \theta_i \cos \theta_i & \sum_{i=1}^n \sin^2 \theta_i \end{bmatrix} \tag{2.37}$$

This relates to the concept of the moment of inertia. In fact, the axis so calculated is identical to that obtained by the method recommended above.

3

Models

3.1 Introduction

Probability models are a very important aspect of statistical analysis. If we can fit a probability model to our data, by suitable estimation of parameters in the model, then the data set can be summarised efficiently using the particular form of probability model specified by the parameter estimates. It is perhaps surprising to find that probability models have not found much application to circular data.

To understand the reason for this last comment, we consider three types of data (linear, circular, spherical) and correspondingly, three models for single groups of data (Normal distribution, von Mises distribution, Fisher distribution); each represents the most commonly used model for its data type. These models have two sorts of parameters, one defining the location or reference direction of the distribution and the other the dispersion about that location. For the Normal distribution, dispersion is quantified by the variance σ^2, with σ^2 near 0 corresponding to a highly concentrated distribution, and with the distribution spreading out more and more over the whole real line as σ^2 increases. For the von Mises and Fisher distributions, the dispersion is quantified by a concentration parameter κ, with $\kappa = 0$ corresponding to uniformity and increasing κ to increasing concentration about the reference direction.

For linear data, the normal distribution is often found to be a satisfactory model, irrespective of the dispersion in the data, and formal statistical analysis can proceed *regardless of the value of* σ^2. For circular or spherical data, formal statistical analysis *cannot* proceed regardless of the value of κ. Roughly speaking, if the underlying von Mises or Fisher distribution has $\kappa \geq 2$, then various statistical methods can be applied because some key approximations are acceptable: $\kappa = 2$ is about the smallest value for which the density at the antipode (see Figure 3.1) has become negligible.

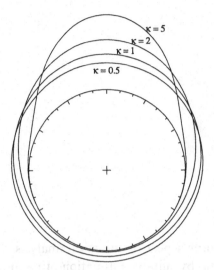

Fig. 3.1 Probability density functions for the von Mises distribution, for different values of the concentration parameter.

When $\kappa < 2$, it is difficult firstly to establish that the model (von Mises or Fisher) fits the data and secondly to apply statistical methods based on the model. Which brings us to the key point: many spherical data sets arising in applications are well-modelled by Fisher distributions with $\kappa \geq 2$, whereas the bulk of circular data sets appear to have underlying distributions with $\kappa < 2$ (if indeed the von Mises model is appropriate).

Despite these remarks, the von Mises distribution is sometimes useful and, indeed, may be the key to making progress on particular problems (for example. with the regression problem of modelling a circular response variable in terms of one or more explanatory variables, as is done in §**6.3**); therefore we shall give a full description of it in this chapter. It is closely approximated by the Wrapped Normal distribution (particularly for $\kappa \geq 2$), so in practice one uses whichever is more convenient. These, plus the circular uniform distribution, are the main models currently in use for analysing circular data. We shall make brief reference only to the host of other circular probability models suggested in the literature. Methods of simulating data from some of these distributions are also given.

3.2 Notation; trigonometric moments

3.2.1 Probability density functions and distribution functions

Let $f(\theta)$ be the *probability density function* (pdf) of a continuous random variable Θ. By this we mean that $f(\theta)$ is a non-negative continuous function such that

$$f(\theta + 2\pi) = f(\theta) \quad (f \text{ is periodic with period } 2\pi) \tag{3.1}$$

and

$$\int_0^{2\pi} f(\theta)\, d\theta = 1 \tag{3.2}$$

The *distribution function* $F(\theta)$ corresponding to $f(\theta)$ can defined over any interval (θ_1, θ_2) by $F(\theta_2) - F(\theta_1) = \int_{\theta_1}^{\theta_2} f(\theta)\, d\theta$ (Mardia 1972a, p. 39). However, it will suffice for our purposes to define the distribution as starting from some convenient point such as $0°$ (or $-180°$), and then define $F(\theta)$ by

$$F(\theta) = \int_0^{\theta} f(\phi)\, d\phi \tag{3.3}$$

In particular,

$$F(2\pi) = 1 \tag{3.4}$$

3.2.2 Trigonometric moments and other population characteristics

The pth *trigonometric moment* of Θ (or of $f(\theta)$), $p = 1, 2, \ldots$, is given by

$$\mu'_p \equiv \rho_p e^{i\mu'_p} = \rho_p \cos \mu'_p + i\rho_p \sin \mu'_p$$
$$= \int_0^{2\pi} \cos(p\theta) f(\theta)\, d\theta + i \int_0^{2\pi} \sin(p\theta) f(\theta)\, d\theta \tag{3.5}$$
$$\equiv \alpha'_p + i\beta'_p \tag{3.6}$$

where α'_p and β'_p are the pth cosine and sine moments respectively.

When $p = 1$, we write simply ρ for ρ_1, μ for μ_1, i.e.

$$\mu'_1 = \rho e^{i\mu}, \tag{3.7}$$

where μ is the *mean direction* and ρ the *mean resultant length*.

The pth *central trigonometric moment* of Θ is just the pth trigonometric moment of $\Theta - \mu$ (Θ centred at its mean direction μ):

$$\mu_p \equiv \rho_p e^{i\mu_p} = \rho_p \cos \mu_p + i\rho_p \sin \mu_p$$
$$= \int_0^{2\pi} \cos p(\theta - \mu) f(\theta)\, d\theta + i \int_0^{2\pi} \sin p(\theta - \mu) f(\theta)\, d\theta \tag{3.8}$$
$$\equiv \alpha_p + i\beta_p$$

When $p = 1$, we get $\alpha_1 = \rho$ and $\beta_1 = 0$, since $\int_0^{2\pi} \cos\theta\, d\theta = \rho \cos\mu$, $\int_0^{2\pi} \sin\theta\, d\theta = \rho \sin\mu$, so that $\int_0^{2\pi} \cos(\theta - \mu)\, d\theta = \rho$ and $\int_0^{2\pi} \sin(\theta - \mu)\, d\theta = 0$. The sample counterparts

$$m'_p, \ m_p, \ \widehat{\rho}_p, \ \widehat{\mu}, \ \widehat{\mu}'_p, \ \bar{C}_p \text{ and } \bar{S}_p$$

of

$$\mu'_p, \ \mu_p, \ \rho_p, \ \mu, \ \mu_p, \ \alpha'_p \text{ and } \beta'_p$$

respectively are described in §**2.3.1**.

Some functions of the first and second trigonometric moments are also of value. The *circular variance* of Θ is defined by

$$v = 1 - \rho, \quad 0 \le v \le 1 \tag{3.9}$$

The *circular standard deviation* is defined, *not* as $\sqrt{v}$ but as

$$\sigma = [-2\log(1-v)]^{\frac{1}{2}} \equiv (-2\log\rho)^{\frac{1}{2}} \tag{3.10}$$

For v small (i.e. ρ near 1)

$$\sigma \simeq (2v)^{\frac{1}{2}} \quad \text{or} \quad [2(1-\rho)]^{\frac{1}{2}} \tag{3.11}$$

the error in the approximation being less than 5% for $v < 0.18$, equivalently $\rho > 0.82$.

Some remarks on the definition and usage of the circular standard deviation are given below, in §**3.3.6**, *Note 2*. An associated measure of spread is the *circular dispersion*

$$\delta = (1 - \rho_2)/(2\rho^2) \tag{3.12}$$

the sample counterpart of which plays an important role in large-sample statistical inference for the mean direction.

Population measures of skewness and kurtosis are defined, respectively, by

$$s = \beta_2/(1-\rho)^{\frac{3}{2}} \tag{3.13}$$

and

$$K = (\alpha_2 - \rho^4)/(1-\rho)^2 \tag{3.14}$$

for populations with non-zero mean resultant length ρ.

Apart from trigonometric moments, there are a number of other population measures of location and spread which can prove useful at times. For a *unimodal* distribution, the population median direction is the unique direction $\tilde{\mu}$ such that

$$\int_{\tilde{\mu}-\pi}^{\tilde{\mu}} f(\theta) \, d\theta = \int_{\tilde{\mu}}^{\tilde{\mu}+\pi} f(\theta) \, d\theta = \tfrac{1}{2} \tag{3.15}$$

where $f(\tilde{\mu}) > f(\tilde{\mu} + \pi)$. If the distribution is *not* unimodal, there is no guarantee that $\tilde{\mu}$ is unique.

The *modal direction* $\breve{\mu}$ is that direction θ for which the pdf $f(\theta)$ is maximised. Again, for multimodal distributions, $\breve{\mu}$ may not be uniquely defined. Turning to measures of dispersion, the *circular mean deviation* is defined by

$$\pi - \int_0^{2\pi} |\pi - |\theta - \tilde{\mu}|| f(\theta) \, d\theta \tag{3.16}$$

(The median direction $\tilde{\mu}$ minimises $\pi - \int_0^{2\pi} |\pi - |\theta - \theta_0|| f(\theta) \, d\theta$ over all choices of θ_0.)

For a unimodal population, it is also useful to have the concepts of *quartiles* of the distribution. The median direction $\tilde{\mu}$ is the 0.5 quantile; the quartiles $\tilde{\mu}_{0.25}$ and $\tilde{\mu}_{0.75}$ are the unique directions each side of $\tilde{\mu}$ such that

$$\int_{\tilde{\mu}-\pi}^{\tilde{\mu}_{0.25}} f(\theta) \, d\theta = \int_{\tilde{\mu}_{0.25}}^{\tilde{\mu}} f(\theta) \, d\theta = \int_{\tilde{\mu}}^{\tilde{\mu}_{0.75}} f(\theta) \, d\theta$$

$$= \int_{\tilde{\mu}_{0.75}}^{\tilde{\mu}+\pi} f(\theta) \, d\theta = 0.25. \tag{3.17}$$

More generally, the p–quantile $\tilde{\mu}_p$ is defined by

$$\int_{\tilde{\mu}-\pi}^{\tilde{\mu}_p} f(\theta) \, d\theta = p, \quad 0 \le p < 1. \tag{3.18}$$

The *inter-quartile range* is simply the angular measure of the arc from $\tilde{\mu}_{0.25}$ to $\tilde{\mu}_{0.75}$ which contains $\tilde{\mu}$.

3.3 Probability distributions on the circle

3.3.1 *The uniform distribution* U_c

With this model, all directions between $0°$ and $360°$ are equally likely.

Probability density function

$$f(\theta) = \frac{1}{2\pi}, \quad 0 \le \theta < 2\pi \tag{3.19}$$

Distribution function

$$F(\theta) = \frac{\theta}{2\pi}, \quad 0 \le \theta < 2\pi. \tag{3.20}$$

Moments

Mean direction μ is undefined

Mean resultant length $\rho = 0$

Circular dispersion $\delta = \infty$

$$\alpha_p = 0, \quad p \geq 1$$
$$\beta_p = 0, \quad p \geq 1$$

Method of simulation As an essential prerequisite to simulation of this and other distributions (and, indeed to use of the bootstrap method throughout this book), we need to have a way of obtaining a large stream of pseudo-random numbers $U_1, U_2, U_3, \ldots$ from $U[0, 1]$, the Uniform distribution on the interval $[0, 1]$. These numbers can be obtained in various ways:

(a) from one's own computer; or
(b) by calling a standard package; or
(c) by programming an algorithm directly.

Method (a) is unreliable, because random number generators on many computers create sequences with considerable pattern, or association, in them. A reliable example of method (b) would be to use the generator in IMSL (IMSL 1991). An example of method (c) is the algorithm of Wichman & Hill (1982) (with the amendment to it by McLeod 1985).

Turning now to the task of simulating an observation Θ from the Circular Uniform distribution, let U be a pseudo-random number from $U[0, 1]$. Then

$$\Theta = 2\pi U$$

Applications The principal value of the uniform model for circular data is as a *null model*, against which various sorts of alternative (unimodal, multimodal) models can be tested.

3.3.2 *The Cardioid Distribution* $C(\mu, \rho)$

This is a symmetric unimodal two-parameter distribution sometimes referred to as the *Cosine* distribution.

Probability density function

$$f(\theta) = \frac{1}{2\pi}\{1 + 2\rho \cos(\theta - \mu)\}, \quad 0 \leq \theta < 2\pi, 0 \leq \rho \leq 1/2 \qquad (3.21)$$

Distribution function

$$F(\theta) = (\rho/\pi) \sin(\theta - \mu) + \theta/(2\pi), \quad 0 \le \theta \le 2\pi \qquad (3.22)$$

Moments

$$\text{Mean direction} \quad \mu$$
$$\text{Mean resultant length} \quad \rho(\le 1/2)$$
$$\text{Circular dispersion} \quad \delta = 1/(2\rho^2)$$
$$\alpha_p = 0, \quad p \ge 2$$
$$\beta_p = 0, \quad p \ge 1$$

Limiting forms As $\rho \to 0$, the distribution converges to the uniform distribution U_c.

Applications The cardioid distribution was introduced by Jeffreys (see Jeffreys, 1961, pp. 328, 330) as follows:

> Consider a circular disk [tray], on which marbles are dropped, while the tray is agitated in its own plane. If the tray is horizontal the chance of a marble coming off is uniformly distributed with regard to the azimuth θ. If it is slightly tilted in the direction $\theta = 0$, the chance will approximate to the above form $\{1 + \alpha f(t)\} \, dt$ with $f(t) = \cos \theta$. But with a larger tilt ...

3.3.3 Wrapped distributions – general

If X is any random variable on the real line, with probability density function $g(x)$, and distribution function $G(x)$, we can obtain a circular random variable Θ by defining

$$\Theta \equiv X[\text{mod } 2\pi]. \qquad (3.23)$$

The probability density function $f(\theta)$ of Θ is obtained by wrapping $g(x)$ around the circumference of a circle of unit radius,

$$f(\theta) = \sum_{k=-\infty}^{\infty} g(\theta + 2k\pi) \qquad (3.24)$$

with corresponding distribution function

$$F(\theta) = \sum_{k=-\infty}^{\infty} [G(\theta + 2\pi k) - G(2\pi k)] \qquad (3.25)$$

A number of useful distributions on the circle can be obtained by this method. For other general information on wrapped distribution, see e.g. Mardia (1972a, §3.4.8).

3.3.4 *The Wrapped Cauchy distribution* $WC(\mu, \rho)$

This is a symmetric unimodal distribution which can be obtained by wrapping the Cauchy distribution (on the line) around the circle.

Probability density function

$$f(\theta) = \frac{1}{2\pi} \frac{1 - \rho^2}{1 + \rho^2 - 2\rho \cos(\theta - \mu)} \qquad 0 \le \theta < 2\pi, \quad 0 \le \rho \le 1$$

(3.26)

Distribution function

$$F(\theta) = \frac{1}{2\pi} \cos^{-1}\left(\frac{(1 + \rho^2)\cos(\theta - \mu) - 2\rho}{1 + \rho^2 - 2\rho \cos(\theta - \mu)}\right) \qquad 0 \le \theta < 2\pi \qquad (3.27)$$

Moments

$$\text{Mean direction} \quad \mu$$
$$\text{Mean resultant length} \quad \rho$$
$$\text{Circular dispersion} \quad \delta = (1 - \rho^2)/(2\rho^2)$$
$$\alpha_p = \rho^p$$
$$\beta_p = 0, \quad p \ge 1.$$

Limiting forms As $\rho \to 0$, the distribution converges to the uniform distribution U_c; as $\rho \to 1$, the distribution tends to the point distribution concentrated in the direction μ.

Method of simulation Let U be a pseudo-random uniform number from $U[0, 1]$ (see §3.3.1), and let $V = \cos(2\pi U)$, $c = 2\rho/(1 + \rho^2)$. Then

$$\Theta = \cos^{-1}\left(\frac{V + c}{1 + cV}\right) + \mu \quad [\text{mod } 2\pi] \tag{3.28}$$

is a pseudo-random value from $WC(\mu, \rho)$.

Applications An indirect application of the Wrapped Cauchy is in simulation of data from a von Mises distribution (cf. §3.3.6). It has also been proposed by Kent & Tyler (1988) as an alternative to the von Mises distribution for modelling symmetric unimodal data; the maximum likelihood estimate of μ represents a robust alternative to the sample mean direction. Kent & Tyler (1988) relate the Wrapped Cauchy distribution to the Angular Central Gaussian distribution in two dimensions by doubling the angles of the latter. See Tyler (1987) and Kent & Tyler (1988) for further details.

For suitably chosen values of the dispersion parameters, the Wrapped Cauchy distribution is quite similar to the Wrapped Normal and von Mises distributions. See §**3.3.6**, *Note 3* for further comments on this.

3.3.5 The Wrapped Normal distribution $WN(\mu, \rho)$

This is a symmetric unimodal two-parameter distribution which can be obtained by wrapping the Normal or Gaussian distribution (on the line) around the circle.

Probability density function

$$f(\theta) = \frac{1}{2\pi}\left(1 + 2\sum_{p=1}^{\infty} \rho^{p^2}\cos p(\theta - \mu)\right), \quad 0 \le \theta < 2\pi, \quad 0 \le \rho \le 1$$

(3.29)

Distribution function This is obtained by straightforward integration of (3.29), but is not required for the purposes of this book.

Moments

$$\text{Mean direction} \quad \mu$$
$$\text{Mean resultant length} \quad \rho$$
$$\text{Circular dispersion} \quad \delta = (1 - \rho^4)/(2\rho^2)$$
$$\alpha_p = \rho^{p^2}$$
$$\beta_p = 0, \quad p \ge 1.$$

If $f(\theta)$ is thought of as being obtained by wrapping a Normal distribution with variance σ^2, then

$$\rho = e^{-\frac{1}{2}\sigma^2}, \quad \text{or} \quad \sigma^2 = -2\log\rho. \tag{3.30}$$

Limiting forms As $\rho \to 0$, the distribution converges to the uniform distribution U_c; as $\rho \to 1$, the distribution tends to the point distribution concentrated in the direction μ.

Method of simulation Carry out the following steps, with the understanding that new pseudo-random values U_1, U_2 are required each time *Step 1* is executed. Note that the algorithm produces a pseudo-random number Z from the Normal $N(0, 1)$ distribution, from which a pseudo-random value Θ from $WN(\mu, \rho)$ is then obtained by wrapping. Calculate the value $\sigma = (-2\log\rho)^{\frac{1}{2}}$, initially.

Step 1.   Let U_1, U_2 be pseudo-random uniform numbers from $U[0, 1]$ (see §**3.3.1**).

Step 2.   Let $z = 1.715528(U_1 - 0.5)/U_2$, $x = 0.25z^2$.

Step 3.   If $x \leq 1 - U_2$ go to *Step 6*.

Step 4.   If $x \leq -\log(U_2)$ go to *Step 6*.

Step 5.   Go to *Step 1*.

Step 6. $Z = z$ is a pseudo-random value from $N(0, 1)$,
$X = \sigma Z + \mu$ is a pseudo-random value from $N(\mu, \sigma^2)$
and
$\Theta = X \,[\text{mod } 2\pi]$ is a pseudo-random value from $WN(\mu, \rho)$.

The algorithm to generate Z is due to Kinderman & Monahan (1977) and Best (1979).

Applications The Wrapped Normal and the von Mises distributions are very similar in terms of shape of distribution, for appropriate choice of dispersion parameters (see §**3.3.6** *Note 3* for further details on this), so in practice one uses whichever is more convenient to the task in hand. Generally, this amounts to using the von Mises model, for which statistical inference is easier. The link to the Normal distribution underlies the definition of circular standard deviation given by (3.10) : see §**3.3.6** *Note 2* for detailed discussion of this point. Use of the Wrapped Normal as a model for geological data dates back at least to Schmidt (1917).

3.3.6   *The von Mises distribution* $VM(\mu, \kappa)$

This is a symmetric unimodal distribution which is the most common model for unimodal samples of circular data.

Probability density function

$$f(\theta) = [2\pi I_0(\kappa)]^{-1} \exp [\kappa \cos (\theta - \mu)] \quad 0 \leq \theta < 2\pi, \quad 0 \leq \kappa < \infty$$

$$(3.31)$$

where

$$I_0(\kappa) = (2\pi)^{-1} \int_0^{2\pi} \exp [\kappa \cos (\phi - \mu)] \, d\phi \qquad (3.32)$$

is the modified Bessel function of order zero; a series expansion and method for evaluating $I_0(\kappa)$ is given in *Note 1(i)* below.

Distribution function There is no simple closed form for this:

$$F(\theta) = [2\pi I_0(\kappa)]^{-1} \int_0^\theta \exp\left[\kappa \cos(\phi - \mu)\right] \mathrm{d}\phi \qquad (3.33)$$

See *Note 3* below, for computational aspects. Batschelet (1981) and Mardia (1972a) give tabulations of the distribution function for a range of values of κ.

Moments

Mean direction $\quad \mu$

Mean resultant length $\quad \rho = A_1(\kappa)$

Circular dispersion $\quad \delta = [\kappa A_1(\kappa)]^{-1}$

$$\alpha_p = A_p(\kappa)$$
$$\beta_p = 0, \quad p \geq 1.$$

where $A_p(\kappa)$ is described in *Note 1* below. Appendices A3 and A4 contain tabulations of ρ as a function of κ, and *vice versa*. Similar tables are also available in Batschelet (1981) and Mardia (1972a).

Limiting forms As $\kappa \to 0$, the distribution converges to the uniform distribution U_c; as $\kappa \to \infty$, the distribution tends to the point distribution concentrated in the direction μ.

Method of simulation Let U_1, U_2 and U_3 be pseudo-random uniform numbers from $U[0, 1]$ (see §3.3.1), and let $a = 1 + (1 + 4\kappa^2)^{\frac{1}{2}}$, $b = [a - (2a)^{\frac{1}{2}}]/(2\kappa)$, $r = (1 + b^2)/(2b)$.

Carry out the following steps, with the understanding that new observations U_1, U_2 or U_3 are used each time step 1, 2 or 4 is executed:

Step 1. Set $z = \cos(\pi U_1)$, $\quad f = (1 + rz)/(r + z)$, $\quad c = \kappa(r - f)$.

Step 2. If $c(2 - c) - U_2 > 0$ go to *Step 4*.

Step 3. If $\log(c/U_2) + 1 - c < 0$ return to *Step 1*.

Step 4. $\Theta = \operatorname{sign}(U_3 - 0.5)\cos^{-1}(f) + \mu \ [\mathrm{mod}\ 2\pi]$.

The algorithm is due to Best & Fisher (1979), and is also available in the IMSL Library (1991). Dagpunar (1990) has suggested a different algorithm which may be faster if κ is changing from call to call.

Applications In many respects, the von Mises distribution is the 'natural' analogue on the circle of the Normal distribution on the real line: see e.g. Mardia (1972a, 1975) for discussion of various similarities. In practice, it does appear to be a reasonable model for symmetric unimodal data sets occurring in many different areas of application; its first use as a model for data is mentioned in Chapter 1. It also serves as a useful statistical model from which to derive procedures of somewhat wider applicability, such as the regression model used in §6.4.

Note 1. Computational details

One difficulty in using the von Mises distribution in practice is the presence of formulae which cannot be evaluated simply. In some cases, adequate approximations are known; more often than not, an iterative procedure or more complex calculation is needed. The difficulties are caused by the presence of modified Bessel functions $I_0(\kappa), I_1(\kappa), \ldots$. Details and further references about these functions beyond what is described here can be found in Mardia (1972a, §3.4.9 and Appendix 1), Batschelet (1981, §16.4), Watson (1983, Appendix A) and Abramowitz & Stegun (1970).

(i) Expressions for $I_p(\kappa), A_p(\kappa)$ and $A_p^{-1}(x)$

$$I_0(\kappa) = \sum_{r=0}^{\infty} (r!)^{-2} (\tfrac{1}{2}\kappa)^{2r} \tag{3.34}$$

and in general,

$$I_p(\kappa) = \sum_{r=0}^{\infty} [(r+p)!r!]^{-1} (\tfrac{1}{2}\kappa)^{2r+p} \quad p = 1, 2, \ldots \tag{3.35}$$

$$A_p(\kappa) = I_p(\kappa)/I_0(\kappa), \quad p = 0, 1, 2, \ldots \tag{3.36}$$

In order to calculate values of $I_p(\kappa)$ and $A_p(\kappa)$ for $p \geq 2$, it suffices to be able to evaluate these functions for $p = 0$ and $p = 1$. Then the following recurrence relationships can be used for $p = 2, 3, \ldots$.

$$I_p(\kappa) = I_{p-2}(\kappa) - \frac{2(p-1)}{\kappa} I_{p-1}(\kappa) \tag{3.37}$$

$$A_p(\kappa) = A_{p-2}(\kappa) - \frac{2(p-1)}{\kappa} A_{p-1}(\kappa) \tag{3.38}$$

In particular,

$$I_2(\kappa) = I_0(\kappa) - \frac{2}{\kappa} I_1(\kappa) \tag{3.39}$$

$$I_3(\kappa) = I_1(\kappa) - \frac{4}{\kappa} I_2(\kappa) \tag{3.40}$$

and

$$A_2(\kappa) = 1 - \frac{2}{\kappa} A_1(\kappa) \tag{3.41}$$

$$A_3(\kappa) = A_1(\kappa) - \frac{4}{\kappa} A_2(\kappa) \tag{3.42}$$

The IMSL Library (1991) has functions which evaluate $I_0(\kappa)$ and $I_1(\kappa)$ to high accuracy: *BSI0* and *DBSI0* are single and double precision versions of $I_0(\kappa)$, respectively, and *BSI1* and *DBSI1* corresponding functions for $I_1(x)$.

Abramowitz & Stegun (1970, p. 378 (§9.8)) give the following polynomial approximations to $I_0(\kappa)$ and $I_1(\kappa)$, from which $A_1(\kappa)$ is readily calculated:

$0 \leq \kappa \leq 3.75, \quad t = \kappa/3.75$

$$
\begin{aligned}
I_0(\kappa) \simeq &1 + 3.5156229t^2 + 3.0899424t^4 + 1.2067492t^6 \\
&+ 0.2659732t^8 + 0.0360768t^{10} + 0.0045813t^{12}
\end{aligned} \tag{3.43}
$$

$$
\begin{aligned}
\kappa^{-1}I_1(\kappa) \simeq &0.5 + 0.87890594t^2 + 0.51498869t^4 \\
&+ 0.15084934t^6 + 0.02658733t^8 \\
&+ 0.00301532t^{10} + 0.00032411t^{12}
\end{aligned} \tag{3.44}
$$

$\kappa \geq 3.75, \quad t = \kappa/3.75$

$$
\begin{aligned}
\kappa^{\frac{1}{2}}e^{-\kappa}I_0(\kappa) \simeq &.39894228 + .01328592t^{-1} \\
&+ 0.00225319t^{-2} - 0.00157565t^{-3} \\
&+ 0.00916281t^{-4} - 0.02057706t^{-5} \\
&+ 0.02635537t^{-6} - 0.01647633t^{-7} \\
&+ 0.00392377t^{-8}
\end{aligned} \tag{3.45}
$$

$$
\begin{aligned}
\kappa^{\frac{1}{2}}e^{-\kappa}I_1(x) \simeq &0.39894228 - 0.03988024t^{-1} \\
&- 0.00362018t^{-2} + 0.00163801t^{-3} \\
&- 0.01031555t^{-4} + 0.02282967t^{-5} \\
&- 0.02895312t^{-6} + 0.01787654t^{-7} \\
&- 0.00420059t^{-8}
\end{aligned} \tag{3.46}
$$

Abramowitz & Stegun (1970, pp. 377–8) give some asymptotic expansions for $I_p(\kappa)$ for $\kappa \to 0$ and $\kappa \to \infty$; see also Mardia (1972a) and Watson (1983) as referenced above.

A simple and reasonably accurate approximation to $A_1^{-1}(x)$ is also available (Best & Fisher, 1981):

$$
A_1^{-1}(x) = \begin{cases} 2x + x^3 + 5x^5/6 & x < 0.53 \\ -0.4 + 1.39x + 0.43/(1 - x) & 0.53 \leq x < 0.85 \\ 1/(x^3 - 4x^2 + 3x) & x \geq 0.85 \end{cases} \tag{3.47}
$$

It is tabulated in Appendix A4.

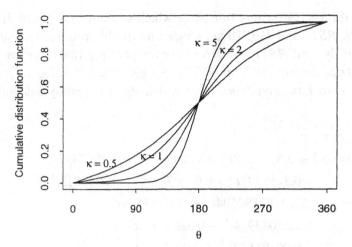

Fig. 3.2 Cumulative distribution functions of von Mises distributions with mean directions 180° and various concentration parameters.

Finally, we give formulae for the first two derivatives of $A_1(\kappa)$:

$$A_1'(\kappa) = \frac{\mathrm{d}A_1(\kappa)}{\mathrm{d}\kappa} = 1 - A_1(\kappa)/\kappa - A_1^2(\kappa) \tag{3.48}$$

$$A_1''(\kappa) = \frac{\mathrm{d}^2 A_1(\kappa)}{\mathrm{d}\kappa^2} = A_1(\kappa)/\kappa^2 - \frac{\mathrm{d}A_1(\kappa)}{\mathrm{d}\kappa} \{2A_1(\kappa) + 1/\kappa\} \tag{3.49}$$

(ii) Expressions for $F_\kappa(\theta) = \{2\pi I_0(\kappa)\}^{-1} \int_0^\theta \exp(\kappa \cos\phi)\,\mathrm{d}\phi$ *and* $F_\kappa^{-1}(p)$

These functions are, respectively, the cumulative distribution function for a von Mises $VM(0, \kappa)$ distribution and the inverse of this function (that is, $F_\kappa^{-1}(P) = \theta$ means that $F_\kappa(\theta) = P$, $0 \le P \le 1$); see Figures 3.2 and 3.3. The inverse function is sometimes referred to as the *quantile* function because $F_\kappa^{-1}(P)$ is the quantile corresponding to a cumulative probability P. One use of the quantile function occurs in carrying out a graphical assessment of the goodness of fit of the von Mises model to a sample of data (cf. §4.5.3(*i*)). This is discussed in more detail after the algorithm is described.

The function

$$I_0(\kappa) F_\kappa(\theta) = (2\pi)^{-1} \int_0^\theta \exp(\kappa \cos\phi)\,\mathrm{d}\phi \tag{3.50}$$

is the incomplete modified Bessel function of order zero. It is a difficult function to evaluate, with different methods being required over different parts of the range of possible κ-values (0 to ∞). A discussion of various methods of approximation is given by Hill (1976), and an implementation of Hill's algorithm is given in Hill (1977).

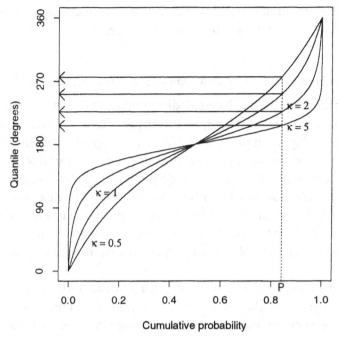

Fig. 3.3 Inverses of the von Mises cumulative distribution functions in Figure 3.2, which give the quantiles $\theta = Q(P)$ of the distribution corresponding to a given cumulative probability P, for distributions with differing concentration parameters.

Assuming that $F_\kappa(\theta)$ can be evaluated to good accuracy, a simple iterative procedure can be used to compute $\theta = F_\kappa^{-1}(P)$. Let ε be some pre-chosen error of approximation of $F_k(\theta)$ to P, e.g. $\varepsilon = 10^{-6}$.

Step 1. $f = 0.5, \quad t = 0, \quad c = \log[I_0(\kappa)]$

Step 2. $g = f - P$
$d = \exp[\log|g| + c - \kappa\cos(t)]$
$t = t - \text{sign}(g)d$
$f = F_\kappa(t)$

Step 3. if $d > \varepsilon$ go to *Step 2*

Step 4. $\theta = t$

An alternative criterion for convergence of the algorithm would be to continue until successive values of $|f - P|$ changed by less than ε (*Step 3*).

For graphical assessment of the goodness-of-fit of the von Mises distribution distribution to a sample of n data points, the algorithm is used to find the quantiles for the closest-fitting von Mises distribution. From §**4.5.3**(i)), the shape of this distribution is $VM(0, \hat{\kappa})$. The sample quantiles are the values $q_i = F_{\hat{\kappa}}^{-1}(P_i)$, where $P_i = i/(n+1), i = 1, \ldots, n$.

Note 2. Circular standard deviation and the von Mises distribution

An important property of the Normal distribution $N(\mu, \sigma^2)$ on the line is that the interval $\mu \pm \sigma$, or indeed $\mu \pm k\sigma$ for any $k > 0$, is an interval of *fixed* probability content (approximately 0.683 for $\mu \pm \sigma$) regardless of the value of σ $(\sigma > 0)$. The reason for this is that a $N(\mu, \sigma^2)$ random variable X can be transformed to a *standardised* $N(0, 1)$ random variable Z by a single change of location and scale : $Z = (X - \mu)/\sigma$. Such is *not* the case with a von Mises $VM(\mu, \kappa)$ random variable; it can be centred (at μ) *but not rescaled to have unit concentration*. At present, there is no circular distribution available with an associated measure of spread, which can be rescaled to have unit spread.

The lack of a location/scale family of circular distributions creates considerable difficulties when the need arises to summarise data sets by a few simple quantities such as mean direction and spread. A notable example of this is in Meteorology: instruments collect large amounts of data on wind speed and direction which then need to be condensed into simple interpretable numbers. Unfortunately, a large amount of historical data on wind directions was summarised by means and standard deviations calculated *as if the data were linear*. Subsequently, much effort has been expended in endeavouring to devise proper circular measures which, in some way, approximate these faulty statistics so as to be able to recover information from the historical summaries. For an account of this problem, see Fisher (1987) and references cited therein. Here, we simply summarise the comments in Fisher (1987) on using $\mu \pm k\sigma$ (σ_c in the reference) as a fixed probability interval around the mean direction μ.

The basic point is that, provided that the von Mises $VM(\mu, \kappa)$ is not too dispersed (i.e. κ is not too small) some reasonable approximations are available. Suppose we desire an interval around μ containing a specified proportion P of the distribution. Let λ_p be a number such that

$$\text{Prob.} (\mu - \lambda_p \sigma_{VM} \leq \Theta \leq \mu + \lambda_p \sigma_{VM}) \simeq P \tag{3.51}$$

where $\sigma_{VM} = [-2 \log A_1(\kappa)]^{\frac{1}{2}}$ is the circular standard deviation for the von Mises distribution. Then some approximate values for λ_p, and their ranges of validity in terms of κ, are:

$$P = 0.50 \quad \lambda_p = 0.66 \quad \kappa \geq 0.24 \quad (\rho \geq 0.12)$$
$$P = 0.75 \quad \lambda_p = 1.12 \quad \kappa \geq 1.12 \quad (\rho \geq 0.17)$$
$$P = 0.90 \quad \lambda_p = 1.69 \quad \kappa \geq 0.65 \quad (\rho \geq 0.31)$$
$$P = 0.95 \quad \lambda_p = 2.06 \quad \kappa \geq 0.80 \quad (\rho \geq 0.37)$$

Fisher (1987) recommended using these λ_p-values for $\rho \geq 0.3$, and then only for $P \leq 0.90$ (misprinted as '$\geq$'). For more dispersed distributions, one is probably best advised to calculate directly a value θ_p such that

$$\text{Prob. } (\mu - \theta_p \leq \Theta \leq \mu + \theta_p) = P \tag{3.52}$$

This can be done using the algorithm in Fisher (1987, Appendix A(b)), which is described below (and corrects the original). Let ε be some desired level of accuracy for θ_p, e.g. $\varepsilon = 0.001$.

Step 0.   Set $a = 2\rho$, $b = \pi P$, $c = \rho^4$, $\varepsilon = 0.001$.
      $i = 0$
      $\psi_i = (b - \pi a)/(1 - a + c)$

Step 1. $f_i = \psi_i + a \sin \psi_i + c \sin (2\psi_i) - b$
      $g_i = 1 + a \cos \psi_i + 2c \cos (2\psi_i)$

Step 2.   Increase i by 1
      $\psi_i = \psi_{i-1} - f_{i-1}/g_{i-1}$

Step 3.   If $|\psi_i - \psi_{i-1}| \geq \varepsilon$ go to *Step 1*

Step 4. $\theta_p = \psi_i$

Note 3. Comparison of the Wrapped Cauchy, Wrapped Normal and von Mises distributions

These three distributions are similar in appearance, and the latter two can be particularly difficult to distinguish as models for data in practical applications. Comparisons between the three have been made by Batschelet (1981, pp. 283–4) and Mardia (1972a, pp. 60–1). Collett & Lewis (1981) noted that sample sizes of the order of a few hundred might be needed to discriminate between the Wrapped Normal and von Mises distributions. So, in practice we use whichever is most convenient. The many optimal and desirable properties of the Normal distribution on the line are, for circular distributions, divided between the Wrapped Normal and the von Mises distribution. In particular, desirable properties relating to statistical inference transfer to the von Mises, which is why it is the model selected for use in this book. See Collett & Lewis (1981) for references to the early work on comparing these families of distributions. There are still other families not described here which are similar in appearance to the Wrapped Normal and von Mises; see Watson (1983, §3.7) for a discussion of the Angular Gaussian and Brownian Motion distributions.

3.3.7 Other models for circular data

As little use is made in this book of the models mentioned below, the reader is referred to an appropriate publication for detailed information.

(i) **Discrete models.** Mardia (1972a, pp. 49–50, 54–5) gives a description of general lattice distributions, including the discrete uniform distribution, and of the Wrapped Poisson distribution.

(ii) **Other symmetric models.** Batschelet (1981 §15.7) discusses some flat-topped and sharply peaked distributions. Mardia (1972a, pp. 51–2) describes a family of triangular distributions.

(iii) **Skew models.** All the distributions described so far are symmetric unimodal models (except when they take the limiting form of uniformity). Batschelet (1981, §15.6) describes a number of asymmetric unimodal models. Mardia (1972a, pp. 52–3) gives a discussion of a distribution arising from a bivariate normal distribution due to Klotz (1964); it is the distribution which arises for wind direction if the x- and y-components of wind velocity are modelled as independent, or more generally correlated, Normal $N(0, \sigma_1^2)$ and $N(0, \sigma_2^2)$ random variables. Mardia (1972a, pp. 51–2) describes a family of skew triangular distributions. These distributions do not seem to have found much application, as yet.

(iv) **Mixtures of distributions.** Suppose that $f(\theta; \mu, \gamma)$ is a unimodal circular probability density function for a random variable Θ with mean direction μ and with the parameter γ controlling the spread of Θ about μ. Let $p_1, \ldots, p_k$ be positive proportions adding to 1; the p_i may be known or unknown. Then the probability density function

$$f^*(\theta) = \sum_{i=1}^{k} p_i f(\theta; \mu_i, \gamma_i) \tag{3.53}$$

is called a mixture distribution. It arises, for example, if several (k) populations are being sampled, and the probability of drawing an observation from population i (with density $f(\theta; \mu_i, \gamma_i)$) is $p_i, i = 1, \ldots, k$. If $\mu_1 = \cdots = \mu_k$, $f^*(\theta)$

is again unimodal. Generally, if some μ_i are unequal, f^* will be multimodal, although in some cases in which the μ_i are sufficiently close together and the dispersions sufficiently large, f^* may be effectively unimodal.

In the smallest mixture case with just $k = 2$ populations, several types of bimodal von Mises distribution have been studied in the literature (see §**4.6**). Generally speaking, mixture models are quite difficult to fit to data (as is also the case on the line) unless the components are reasonably distinct.

(**v**) **Bivariate models: circular–linear.** Several models have been proposed for the joint distribution of circular and linear random variable, usually in conjunction with a parameter which measures some form of circular–linear association. These can be found in Johnson & Wehrly (1978), Jupp & Mardia (1980) and references therein.

(**vi**) **Bivariate models: circular–circular.** The same general comment applies as in *(v)*; see, for example Wehrly & Johnson (1979), Jupp & Mardia (1980), Rivest (1982) and Fisher & Lee (1983).

4

Analysis of a single sample of data

4.1 Introduction

The methods in this chapter are concerned with the analysis of a sample of independent observations $\theta_1, ..., \theta_n$ from some common population of vectors or axes. A number of data sets were introduced in Chapter 2, in the context of methods for displaying data. Here, we begin with some other examples to illustrate the range of problems which occur with a single sample of data.

Example 4.1 Figure 4.1 shows a plot of the directional preferences of 50 starhead top-minnows, after they had been displaced to unfamiliar surroundings and then subjected to heavily overcast conditions. (The data are listed in Appendix B4.) The fish use a sun compass to move in a direction which, at the location of their capture, would return them to the land–water interface. It is of interest to know whether they still have a preferred orientation, or whether the distribution is essentially uniform on $(0°, 360°)$.

Example 4.2 Figure 4.2 shows a plot of measurements of long-axis orientation of 60 feldspar laths in basalt. (Both ends of each axis are plotted. The data are listed in Appendix B5.) In this example, it is of interest to know whether the distribution is essentially uniform, or whether there are, in fact, several modal groups present.

Example 4.3 Figure 4.3 shows a plot of 30 cross-bed azimuths of palaeocurrents measured in the Belford Anticline in New South Wales. (The data are listed as the second set in Appendix B6.) The sample appears to have a single modal group. It is desired to summarise the data with a few sample statistics, which can then be compared and possibly combined with corresponding information from other samples in the region.

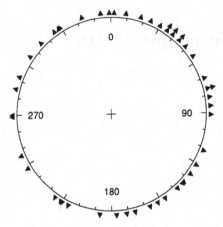

Fig. 4.1 Sun compass directions of 50 starhead minnows, measured under overcast conditions. See Example 4.1.

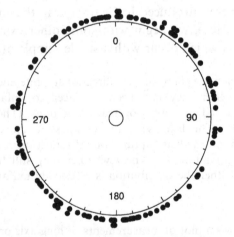

Fig. 4.2 Measurements of the long-axis orientations of 60 feldspar laths in basalt. See Example 4.2.

Example 4.4 Figure 4.4 shows a plot of the directions chosen by 100 ants in response to an evenly illuminated black target placed as shown. The ants tended to run toward the target. In order to carry out statistical inference as efficiently as possible, it is of interest to ascertain whether the data can be modelled as a random sample from a von Mises distribution. (The data are listed in Appendix B7, as part of a larger sample from which they were randomly selected.)

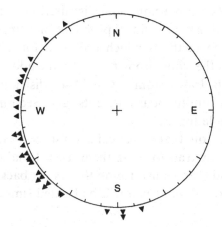

Fig. 4.3 Measurements of 30 cross-bed palaeocurrents in the Belford Anticline, New South Wales. See Example 4.3.

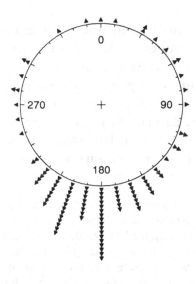

Fig. 4.4 Directions chosen by 100 ants in response to an evenly illuminated black target placed as shown. See Example 4.4.

In the *exploratory* phase of statistical analysis, we consider simple questions, such as whether the data appear to be uniform, unimodal or multimodal, and whether there are any unusual features such as outliers. Such general questions are considered in Section 4.2. Sections 4.3–4.5 are mostly concerned with analysing vectorial or axial data from a unimodal distribution. Section 4.3 considers tests of uniformity. Section 4.4 considers problems of estimation and testing for unimodal data, using methods which are valid for a wide range of data sets; Section 4.5 treats the same problems under the additional assumption that the data have been drawn from a von Mises distribution. Section 4.6 contains a discussion of multimodal data sets, and Section 4.7 some references to other single-sample topics.

As a final but important point, we note that statistical analysis of a sample of *axial* data will generally be done by transforming them to vectorial data, using methods for vectorial data, and then transforming the results back. So, there is no separate discussion of axial data, which will be treated simply by example.

4.2 Exploratory analysis

This is the initial phase of statistical analysis, in which the methods of §**2.2** are used to highlight qualitative features of the sample of data such as uniformity, unimodality, multimodality, symmetry, outlying points. The main methods at our disposal are the *raw* data plot and the nonparametric density estimate, examples of which have already been seen in §**2.2**(*iv*). Nonparametric density estimates can be particularly helpful in investigating the distributional shape in a large data set with data values spread right around the circle. As in Example 2.14, we use the superior ability of the linear version to highlight modes, in combination with the circular display.

> **Example 4.5** For the axial data of Example 4.2 displayed in Figure 4.2, two sets of nonparametric density estimates are shown in Figure 4.5, based on smoothing values $h = 0.5$ and 0.75 applied to the data when converted to vectors (cf. §**2.2**(*iv*)). Because the data are axial, plotting the density over $(0, 360°)$ suffices to give us two complete cycles of the density. As the amount of smoothing increases, we see some of the small bumps in Figures 4.5(a) and its linear companion 4.5(b) disappearing, leaving the possibility of one major group, and maybe one or two minor groups. This will be investigated further (Examples 4.7, 4.8, and 4.11) with graphical and formal tests.

Occasionally, we may want to assume that the sample is drawn from a unimodal distribution which is symmetric about its median direction, to allow us to use simplified and more efficient statistical methods. The validity of this assumption can be explored by a simple graphical method.

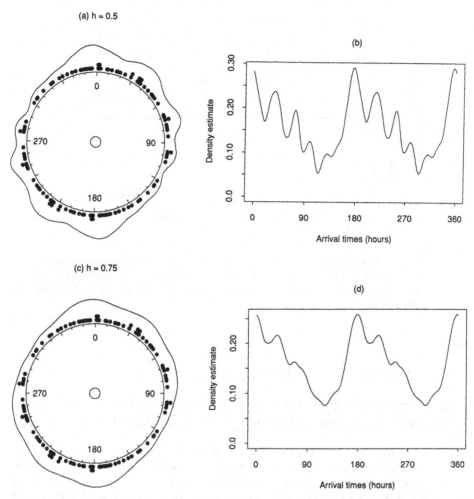

Fig. 4.5 Effect of increasing the smoothing parameter h in a nonparametric density estimate, for the measurements of long-axis orientations shown in Figure 4.2. As the amount of smoothing increases, many small bumps disappear, leaving the possibility of one major modal group and perhaps one or two minor groups. (a) $h = 0.5$. (b) linear version of (a). (c) $h = 0.75$. (d) linear version of (b). See Example 4.5.

Calculate the median vector $\widetilde{\theta}$ (cf. §2.3.2), and then the values

$$z_i = \tfrac{1}{2}(\theta_i - \widetilde{\theta}), \quad -\tfrac{1}{2}\pi \le z_i \le \tfrac{1}{2}\pi, \quad i = 1, \ldots, n \qquad (4.1)$$

and rearrange them into ascending order to get

$$z_{(1)} \le \cdots \le z_{(n)} \qquad (4.2)$$

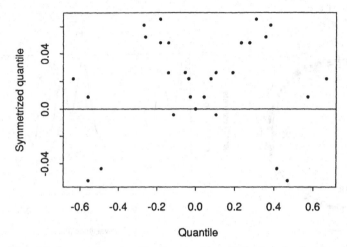

Fig. 4.6 Symmetry plot for the sample of 30 cross-bed palaeocurrents shown in Figure 4.3. The systematic way in which the points depart from the *x*-axis suggests that the assumption of symmetry of the distribution may be dubious. See Example 4.6.

Plot

$$(\sin z_{(i)}, \sin(\tfrac{1}{2}(z_{(i)} + z_{(n+1-i)}))), \quad i = 1, \ldots, n \tag{4.3}$$

If the underlying distribution is symmetric about its median direction $\tilde{\theta}_0$, the points should plot approximately along a horizontal line close to the *x*-axis.

Example 4.6 For the palaeocurrent data of Example 4.3 plotted in Figure 4.3, a symmetry plot is shown in Figure 4.6. Although close to the horizontal line $x = 0$ the points seem to depart in a patterned manner from this line, and a formal test of symmetry may be advisable: see Example 4.17.

Note: with a large sample of very dispersed data, some detail can be lost at the extreme ends of the plot. In such cases, it may be helpful to redefine the values z_i as

$$z_i = \tfrac{1}{4}(\theta_i - \tilde{\theta}) \tag{4.4}$$

4.3 Testing a sample of unit vectors for uniformity

Uniformity (also referred to as *randomness* or *isotropy*) of a sample of circular measurements can be a difficult property to assess by eye. Not infrequently, seemingly clustered data turn out to be manifestations of a uniform distribution when subjected to a formal statistical test. Conversely, a seemingly uniform data set may prove to have a significant concentration in a single (possibly specified) direction.

It is precisely this latter problem that makes the choice of uniformity test – and such tests are legion – crucially dependent on the alternative model to uniformity. If we are simply interested in detecting *any* sort of non-uniform alternative model (unimodal, bimodal, trimodal ...) then we select a so-called *omnibus* test. An omnibus test is capable of detecting any departure from uniformity, given a sufficiently large sample. However, because it has to be able to detect *all* alternatives, it may well not be particularly effective in detecting some particular alternative of interest, such as unimodality. If we are specifically interested in the possibility that the distribution has a single mode then we should use a test designed for this purpose. Note, however, that a specific test such as this may be useless against other types of alternative model.

(i) Graphical assessment of uniformity

Before carrying out any sort of formal test, it is useful to make a uniform probability plot of the data. Calculate the linear order statistics $\theta_{(1)} \leq \cdots \leq \theta_{(n)}$ of the sample as described in §**2.3.3**, and let

$$x_1 = \theta_{(1)}/2\pi, \ \ldots, \ x_n = \theta_{(n)}/2\pi \qquad (4.5)$$

Now plot the values

$$[1/(n+1), x_1], \ldots, [n/(n+1), x_n] \qquad (4.6)$$

The points should lie roughly along a 45° line passing through (0, 0), if the uniform model is correct.

Because the data are circular, there is an arbitrary choice for the beginning of the plot. Any artefact due to this choice can be avoided by taking the first 20% (say) of data pairs in (4.6), adding 1 to each value in each pair, and plotting these points; similarly, take the last 20% of data pairs, subtract 1 from each value in each pair, and plot them; this brings in the first few points at the lower end (say from 0.0 to 0.2) as an extension at the upper end (from 1.0 to 1.2), and with a similar extension at the lower end.

> **Example 4.7** For the axial data considered earlier in Examples 4.2 and 4.5, to assess possible departure from a model of uniformity, we convert the sample values to vectors by doubling them (and reducing them modulo 360°), and then proceed as described above, to get Figure 4.7(a). There is some evidence of a departure from the uniform model, because of the lengthy excursion of the plotted points below the 45° line between 0.0 and

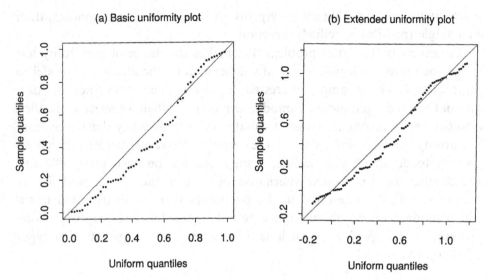

Fig. 4.7 Uniformity plots for the measurements of long-axis orientations displayed in Figure 4.2. Departure from the model of uniformity, or isotropy, is suggested by deviations from the 45°-line shown on the graph. (a) Basic uniformity plot. (b) Uniformity plot in (a) extended at each end by partial wrap-around. There is some evidence of such a departure in the plot. See Example 4.7.

0.7, suggesting a single large modal group. An enhanced version of the graph, with the additional values

$$\left(\frac{1}{61} + 1, x_1 + 1\right), \dots, \left(\frac{10}{61} + 1, x_{10} + 1\right), \tag{4.7}$$

and

$$\left(\frac{51}{61} - 1, x_{51} - 1\right), \dots, \left(\frac{60}{61} + 1, x_{60} - 1\right) \tag{4.8}$$

plotted, is shown in Figure 4.7(b). A test of uniformity against a bipolar distribution (an axial distribution with a single modal group) is considered below.

(ii) Test of randomness against any alternative

It will often be desirable to supplement the graphical procedure with a formal test. Following on from the earlier discussion, we shall consider two types of tests, beginning with the omnibus type. We also need to distinguish between continuous (that is, *ungrouped*) data and grouped data. Suppose first that the data are ungrouped. Calculate the values $x_1, \dots, x_n$ as in (4.5), and then the statistics

$$D_n^+ = \text{maximum of} \quad \frac{1}{n} - x_1, \ \frac{2}{n} - x_2, \ \dots, 1 - x_n, \tag{4.9}$$

$$D_n^- = \text{maximum of} \quad x_1, \ x_2 - \frac{1}{n}, \ x_3 - \frac{2}{n}, \ \dots, \ x_n - \frac{n-1}{n} \tag{4.10}$$

$$V_n = D_n^+ + D_n^- \tag{4.11}$$

and

$$V = V_n(n^{\frac{1}{2}} + 0.155 + 0.24/n^{\frac{1}{2}}) \tag{4.12}$$

The hypothesis that the sample $\theta_1, \dots, \theta_n$ was drawn from a uniform distribution is rejected if V is too large.

Critical values. See Appendix A5 for a table of critical values. Significance probabilities, or P-values, of V_n have been tabulated by Arsham (1988).

> **Example 4.8** For the data analysed in Examples 4.2, 4.5 and 4.7, we obtain (after converting the axes to vectors) $D_n^+ = 0.1389, D_n^- = 0.0611$, and $V = 1.586$. Since this value of V only just exceeds the upper 15% point in Appendix A5, we conclude that there is little evidence for a departure from uniformity, based on this test. Bear in mind, however, that we have tested the null hypothesis of uniformity against *all possible* alternative models. A more specific test can produce a different result, as we shall see in Example 4.11.

We now turn to the case of *grouped* data. In some areas of application, it is not uncommon that the data are recorded in say, 5° or 10° intervals. In such cases, the above test is inappropriate; instead, a chi-squared test can be used. Suppose that the recording intervals are all equal to ϕ_0°, giving k intervals (where $k \times \phi_0^\circ = 360°$), and let $n_1, \dots, n_k$ ($n_1 + \dots + n_k = n$) be the observed numbers of data in these k intervals. If the underlying distribution *is* uniform, we *should* get about $m = n/k$ in each interval. As a rule, m should be at least 2, or equivalently,

$$\phi_0^\circ \le 2 \times 360°/n$$

If this is not the case, the number of intervals should be reduced by pooling pairs of intervals (e.g. 5° grouping goes to 10° grouping, hence halving the number of intervals k) until this rule is satisfied.

Suppose that the data are already grouped satisfactorily. Calculate

$$Y = (k/n) \sum_{i=1}^{k} n_i^2 - n \tag{4.13}$$

The hypothesis of uniformity is rejected if Y is too large.

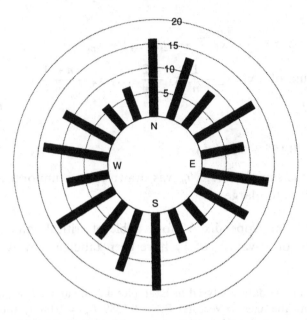

Fig. 4.8 Angular histogram of the horizontal axes of 100 outwash pebbles from a late Wisconsin outwash terrace. The grouping interval is 20°. See Example 4.9.

Critical values. Reject the hypothesis at level α if Y exceeds the upper $100\alpha\%$ value of the χ^2_{k-1} distribution in Appendix A2.

A useful alternative form of (4.13) is obtained by calculating the difference $d_i = n_i - m$ between the observed frequency of data in each interval i and the expected frequency m. Then we can write

$$Y = (1/m) \sum_{i=1}^{k} d_i^2 \tag{4.14}$$

The pattern of signs of the sequence $d_1, \ldots, d_n$ (particularly for large absolute deviations $|d_i|$) could then be used to suggest particular departures from the Uniform model: e.g. the pattern

$$- \ - \ + + + - \ - \ - \ - \ - \ + + + +$$

would suggest a bimodal distribution.

Example 4.9 Figure 4.8 is an angular histogram of the horizontal axes of 100 outwash pebbles from a late Wisconsin outwash terrace. (The data are listed in Appendix B8.) They were recorded in 20° intervals, so that to test the hypothesis of uniformity against a general alternative hypothesis (unimodality or multimodality), we need to use a chi-squared test because of the substantial grouping. There are $k = 9$ grouping intervals (*not* 18, because the data are axial, not vectorial), from which we calculate $Y = 8.72$, which is much less than the upper 20% point of the χ_8^2 distribution. Alternatively, using the formula for the quantile function of the χ^2 distribution in Appendix A2, the significance probability of Y is 0.63.)

(iii) Test of randomness against a unimodal alternative

Now suppose that we are interested in detecting a single modal direction in a sample of vectors (or, for axes, a bipolar sample which converts to a unimodal sample when the axes are converted to vectors). A different test of uniformity, known as the Rayleigh test, is appropriate for this purpose.

(a) Specified mean direction If the mean direction of the alternative unimodal model is specified as μ_0, calculate $\bar{\theta}$ and $\bar{R}$ as in (2.8) and (2.9), and then

$$\bar{R}_0 = \bar{R}\cos(\bar{\theta} - \mu_0) \tag{4.15}$$

The null hypothesis of uniformity is rejected if $\bar{R}_0$ is too large.

Critical values. Set $Z_0 = (2n)^{\frac{1}{2}}\bar{R}_0$, and find $p_z = \Phi(Z_0)$ from Appendix A1. Let $f_z = \exp(-\frac{1}{2}z_0^2)/(2\pi)^{\frac{1}{2}}$, and calculate

$$P_0 = 1 - p_z + f_z[(3Z_0 - Z_0^3)/(16n)$$
$$+ (15Z_0 + 305Z_0^3 - 125Z_0^5 + 9Z_0^7)/(4608n^2)] \tag{4.16}$$

Then P_0 is the significance probability of $\bar{R}_0$.

Example 4.10 Figure 4.9 shows the dance directions of 279 honey bees viewing a zenith patch of artificially polarised light, measured as part of an experiment to demonstrate that specialised photoreceptors at the dorsal margin of the eye are necessary for detecting polarised skylight and deriving compass information from celestial *e*-vector patterns. (The data are listed in Appendix B9.) In this control sample, that part of the eye of each bee conjectured to contain the receptors had been painted out. So it is of interest to test the null hypothesis of uniformity against the alternative of a preferred N–S axis for the dance directions. Because of the non-directional nature of the alternative hypothesis, we can handle the problem by treating the data as axes, converting them to vectors, and then testing the hypothesis that $\mu = 0$. (Thus, double all the data and then reduce them modulo 360°.) We get $\bar{R}_0 = 0.0472$, whence $Z_0 = 1.12$ and a significance probability $P_0 = 0.867$ from (4.16). So there is no evidence that the bees are responding to the polarised light in terms of preferred dance direction.

It is worth noting in this example that using the omnibus test in §**4.3**(*ii*) yields $V = 1.87$ which, from Appendix A5, lies between the 5% and 1% points, so there is some evidence of a departure from uniformity using this

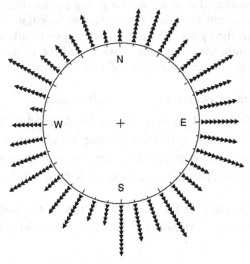

Fig. 4.9 Dance directions of 279 honey bees viewing a zenith patch of artificially polarised light. It is of interest to determine whether the bees' dances are in random directions, or whether the dances have a preferred North–South orientation. See Example 4.10.

test, but in the direction of multimodality rather than unimodality in view of the analysis above. The converse effect can also be observed, as in the case of Example 4.11 below.

(b) Unspecified mean direction. More commonly, μ_0 is not specified. In this case, calculate $\bar{R}$ as in (2.10). The null hypothesis of uniformity is rejected if $\bar{R}$ is too large.

Critical values. Calculate $Z = n\bar{R}^2$, and hence the significance probability P of $\bar{R}$ as

$$P = \exp(-Z)[1 + (2Z - Z^2)/(4n)$$
$$- (24Z - 132Z^2 + 76Z^3 - 9Z^4)/(288n^2)] \qquad (4.17)$$

For $n \geq 50$, the approximation

$$P = \exp(-Z) \qquad (4.18)$$

is quite adequate.

Example 4.11 Continuing the analysis of the data in Figure 4.2 (cf. Examples 4.2, 4.5, 4.7 and 4.8), if we test the null hypothesis of uniformity against the alternative of a bipolar distribution (unimodal axial distribution) using the Rayleigh statistic, after converting the axes to vectors we get $\bar{R} = 0.2370$, whence $P = \exp(-n\bar{R}^2) = 0.034$. So there is some evidence for the presence of a mode in the underlying distribution. By way of contrast, note that when a less specific test was used (Example 4.8), there was little evidence to suggest a departure from the null model of uniformity.

As with the test based on $\bar{R}_0$, if the data are grouped rather than continuous, the same tests can be applied with each data point being given the value of the midpoint of its grouping interval.

> **Example 4.12** For the sun compass orientations of top-minnows described in Example 4.1, the interesting alternative hypothesis to uniformity is that the minnows have a single preferred direction. We obtain $\bar{R} = 0.1798$, and $Y = 1.617$, leading to a significance probability of 0.20, so there is little evidence that the minnows can orient themselves on heavily overcast days.

References and footnotes. The test based on V is due to Kuiper (1960), who derived its asymptotic distribution. Its small-sample distribution was given in Stephens (1965) and the succinct tabulation in Appendix A5 in Stephens (1970). The tests based on $\bar{R}$ and $\bar{R}_0$ are due to Rayleigh (1880, 1905, 1919); see also Chapter 1. Another common omnibus test is the so-called U^2 statistic (Watson, 1961; see also Mardia, 1972a pp. 180–2; and Batschelet, 1981 pp. 79–81), which is used later in this chapter (§**4.5.3**(*ii*)) as a goodness-of-fit statistic for the von Mises distribution. The Rayleigh test and Watson's U^2 are members of a general class of tests due to Beran (1968, 1969a). For information on these and other tests, particularly those based on spacings, see e.g. Mardia (1972a §7.2), Batschelet (1981, Chapter 4); Jammalamadaka (J. S. Rao) (1984) and Upton & Fingleton (1989, Chapter 9). Rao (1969) discussed variants of chi-squared tests for testing uniformity on the circle. Another test for uniformity which is designed to have good power against both unimodal and bimodal alternatives is due to Hermans & Rasson (1985); at present it is only tabulated for sample sizes 10, 20 and ∞ (i.e. very large n) and for three critical levels. Moore (1980) gave a modification of the Rayleigh test for the case when each unit vector has an associated length. The approximations to the significance probabilities of Z_0 and Z are due to Greenwood and Durand (1955); see also Mardia (1972a, pp. 133, 135). A test of uniformity against *sectoral* preference – that is, that the data tend to cluster in a specified arc of the circle – has been given by Lehmacher & Lienert (1980). For the case of grouped data, Freedman (1979, 1981) proposed two statistics, one a modification of the Kuiper statistic V and the other a modification of the Watson statistic U^2.

4.4 Nonparametric methods for unimodal data

4.4.1 Introduction

Having decided that a given sample of data manifests evidence of at least one modal group, we turn to methods for investigating a sample of unimodal data or a unimodal subset of a larger data set. The problems of interest relate to estimating a 'preferred direction' (together with a suitable error of estimate) and to testing whether the sample 'preferred direction' accords with some specified value. 'Preferred direction' will be taken to mean either the sample median direction or the sample mean direction (§**2.3.1** and §**2.3.2**), and we consider methods based on each, separately.

For a sample of *linear* data, the median and the mean can take widely differing values due to the way each is influenced by outlying values. Whilst this is true in principle for circular data, it is far less common in practice. With linear data, the tails of the distribution may extend indefinitely far out, but this is clearly not true of circular data. Of course, a circular data set may be concentrated in just a few degrees of arc, except for a few aberrant points, thus mimicking the 'long-tailed' phenomenon for linear data. However, many unimodal circular data sets encountered in practice are sufficiently dispersed that discordant observations may be difficult to spot, let alone have significant influence on estimation of the mean direction. Generally, the sample mean direction is to be preferred, particularly in moderate to large samples, partly because it is far easier to calculate, but more importantly because it can be combined with a measure of sample dispersion to act as a summary of the data suitable for comparison and amalgamation with other such information.

Notwithstanding these remarks, the sample median direction has its applications (see §**7.2.3** for example), and affords simply-computed exact confidence intervals for the population median direction in small samples.

4.4.2 Estimation of the median direction

The point estimate $\tilde{\theta}$ of $\tilde{\mu}$ is calculated as in §**2.3.2**. A confidence region for $\tilde{\mu}$ can be found by analogy with methods for linear data; the method assumes that the data are not too dispersed (e.g. clearly concentrated on an arc substantially less than the whole circumference).

For any integer $m = 1, 2, \ldots$, count off m θ-values to the left and to the right of $\tilde{\theta}$ (not counting $\tilde{\theta}$ itself) to get lower and upper data values $\theta_{(L_m)}$, $\theta_{(U_m)}$ say. The interval $(\theta_{(L_m)}, \theta_{(U_m)})$ containing $\tilde{\theta}$ will then be a confidence interval for $\tilde{\mu}$ whose confidence level α depends on m. For small samples, only a very limited range of α-levels is available, but this problem disappears with increasing sample size n.

(a) $n < 16$. Appendix A6 lists a range of values m and corresponding exact α–levels for the confidence interval $(\theta_{(L_m)}, \theta_{(U_m)})$.

(b) $n \geq 16$. An approximate $100(1 - \alpha)\%$ confidence interval can be found by setting

$$m = 1 + \text{integer part of} \quad \tfrac{1}{2}n^{\frac{1}{2}}z_{\frac{1}{2}\alpha} \tag{4.19}$$

where $z_{\frac{1}{2}\alpha}$ is the upper $100(\tfrac{1}{2}\alpha)\%$ of the $N(0, 1)$ distribution in Appendix A1.

Example 4.13 The three plots in Figure 4.10 are of the first 10, the first 20, and the entire data set of Figure 4.3. We shall calculate point estimates and confidence intervals (of approximate size 95%) for the population median direction using the two subsamples. The sample median directions of these subsamples are $279°$ ($n = 10$) and $247.5°$ ($n = 20$). For $n = 10$, a confidence interval of exact size 97.8% can be computed using Appendix A6: this results in the interval $(245°, 315°)$. For $n = 20$, the approximate method can be used. For $\alpha = 0.05$, $z_{\frac{1}{2}\alpha} = 1.96$, $m = $ integer part of $[\frac{1}{2}(20)^{\frac{1}{2}} \times 1.96] = 4$. So, counting off five observations in the ordered data set each side of the sample median $247.5°$ gives an approximate 95% confidence interval $(229°, 277°)$.

References and footnotes. The approximate confidence interval in (4.17) is based on the same rule for linear data, due to MacKinnon (1964) (see also David, 1981).

4.4.3 *Testing the median direction for a specified value*

Suppose we wish to test the hypothesis that $\tilde{\mu}$ takes the particular value $\tilde{\mu}_0$, against the alternative that $\tilde{\mu} \neq \tilde{\mu}_0$, at some level α. If some sample values actually equal $\tilde{\mu}_0$, denote the number of such values by k. We can handle the testing problem in two ways (at least, for $n - k \geq 16$).

Firstly, if we have already calculated a $100(1 - \alpha)\%$ confidence interval for $\tilde{\mu}$, we can see whether or not $\tilde{\mu}_0$ falls inside this confidence interval. Alternatively, we can carry out a direct test (presented here for $n \geq 16$; for $n < 16$, use the confidence interval method). Let m be the number of data points in the arc $(\tilde{\mu}_0, \tilde{\mu}_0 + 180°)$ not equal to $\tilde{\mu}_0$. If $\tilde{\mu}_0$ is the true median direction, m should not differ too much from $(n - k)/2$.

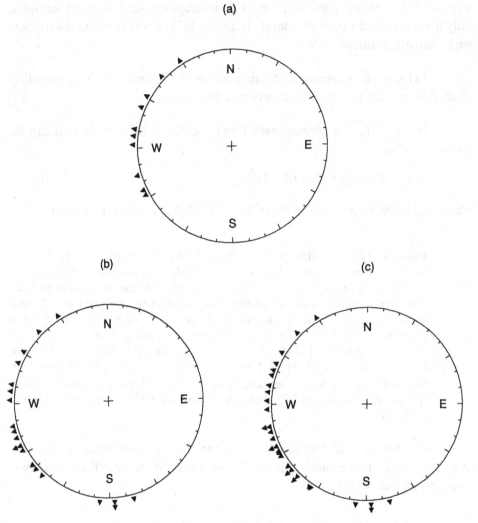

Fig. 4.10 Two subsamples of the 30 cross-bed palaeocurrents shown in Figure 4.3, together with the complete sample. (a) Subsample of size 10. (b) Subsample of size 20. (c) Complete sample.

Let

$$Y = (2m - n + k)^2/(n - k) \qquad (4.20)$$

The hypothesis that $\tilde{\mu} = \tilde{\mu}_0$ is rejected if Y is too large.

Critical values. For a $100\alpha\%$ test, compare Y with the upper $100\alpha\%$ point of the χ_1^2 distribution in Appendix A2.

4.4.4 Estimation of the mean direction

The point estimate $\hat{\mu}$ of μ is $\bar{\theta}$, given by (2.9). To calculate an interval estimate, we need two different approaches, depending on sample size.

(a) $n < 25$. A confidence interval for μ can be obtained using bootstrap methods (see §8). The basic mathematical calculations are given in §8.3. The only decision to make before commencing is whether or not the sample seems to have been drawn from a symmetric distribution, as this will affect how we generate our bootstrap samples. A test for symmetry is given in §4.4.6.

Note that, as an alternative to calculating a confidence interval for μ, we can display a nonparametric density estimate of the distribution of $\hat{\mu}$ computed from the bootstrap values: see *Stage 4, Technique 3* in §8.3.2.

> **Example 4.14** For the two subsamples of Example 4.13, the mean directions are $280.0°$ ($n = 10$) and $248.7°$ ($n = 20$). Using 200 bootstrap resamples, we get approximate 95% confidence intervals for the unknown mean direction as $(256.6°, 313.5°)$ based on the sample of size 10, and $(213.9°, 268.6°)$ based on the sample of 20. (*Technique 1* in §8.3.2 *Stage 3* has been used for these confidence intervals.) These can be compared with the 95% confidence interval for the full data set calculated in Example 4.16.
>
> The first subsample is sufficiently small that we might consider using the iterated bootstrap method (§8.3.3) to get a confidence interval with coverage closer to 95%. Using $B = 200$, we get the first few p_b values and the last few q_b values in equations (8.15) and (8.16) to be
>
> $$0.000 \ 0.000 \ 0.000 \ 0.005 \ 0.005 \ 0.010 \ 0.010 \ 0.010 \ 0.015 \ 0.020 \ 0.020$$
> $$0.025 \ 0.025 \ 0.030 \ 0.030 \ 0.030 \ 0.035 \ 0.035 \ldots$$
>
> and
>
> $$\ldots 0.050 \ 0.050 \ 0.050 \ 0.050 \ 0.045 \ 0.045 \ 0.030 \ 0.030 \ 0.020 \ 0.020$$
> $$0.020 \ 0.020 \ 0.020 \ 0.015 \ 0.010$$
>
> respectively; whereas with the un-iterated bootstrap, these would simply be
>
> $$\frac{1}{B} = 0.005, \frac{2}{B} = 0.010, \frac{3}{B} = 0.015, \ldots$$
>
> and
>
> $$\ldots, \frac{B-2}{B} = 0.990, \frac{B-1}{B} = 0.995, \frac{B}{B} = 1.000$$
>
> This leads to the calibrated 95% bootstrap confidence interval of $(262.6°, 312.4°)$, a few degrees shorter than the uncalibrated interval.
>
> **Example 4.15** The data in Figure 4.11 show the directions of 11 long-legged desert ants (*Cataglyphis fortis*) after one eye on each ant was 'trained' to

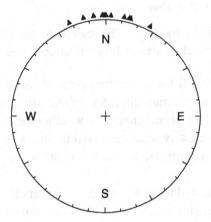

Fig. 4.11 Directions of 11 ants after they had learned their home direction (here North) using one eye, and then had that eye occluded. See Example 4.15.

learn the ant's home direction, then covered and the other eye uncovered. (The data are the first set in Appendix B10.) Home direction is 0° for each ant. The estimate of their mean direction is 2.3°. To calculate a confidence region for the mean direction, we shall assume that the data are from a symmetric distribution. Using $B = 200$ resamples, we calculate the 95% confidence region as $2.3° \pm 13.5°$, from §**8.3.2** *Stage 4, Technique 2*.

(b) $n \geq 25$. For moderate to large samples, we can avail ourselves of simpler and more flexible methods. Calculate the circular dispersion $\widehat{\delta}$ as in (2.28), and the *circular standard error* $\widehat{\sigma}$ of $\widehat{\mu}$ from

$$\widehat{\sigma}^2 = \widehat{\delta}/n \qquad (4.21)$$

Then an approximate $100(1 - \alpha)\%$ confidence interval for μ is given by

$$(\widehat{\mu} - \sin^{-1}(z_{\frac{1}{2}\alpha}\widehat{\sigma}), \quad \widehat{\mu} + \sin^{-1}(z_{\frac{1}{2}\alpha}\widehat{\sigma})) \qquad (4.22)$$

where $z_{\frac{1}{2}\alpha}$ is the upper $100(\frac{1}{2}\alpha)\%$ point of the $N(0, 1)$ distribution (Appendix A1).

Example 4.16 Continuing the analysis of the 30 cross-bed azimuths in Example 4.3 (cf. Example 4.13), we get the estimated mean direction as 247.6°, $\widehat{\sigma} = 0.1314$, and an approximate 95% confidence region for the mean direction as $247.6° \pm 14.9°$, or $(232.7°, 262.5°)$. The sets of confidence regions for the two subsamples and for the full sample are shown in Table 4.1. These results, and the additional entries in Table 4.1, are based

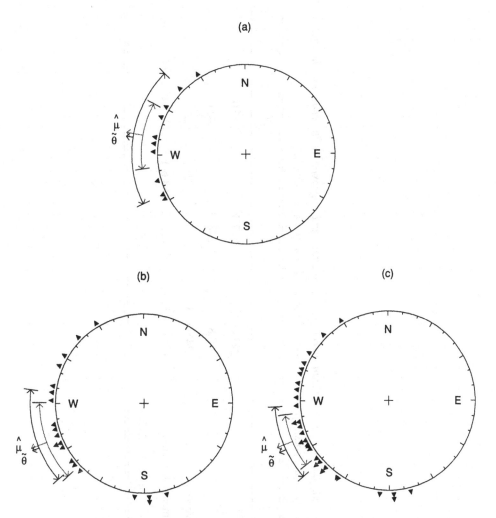

Fig. 4.12 The three data sets of cross-bed palaeocurrents shown in Figure 4.10, augmented with the summary information in Table 4.1. The median and mean directions are shown for each sample, as is the associated 95% confidence interval (dashed lines relate to mean directions). For the two smaller sample sizes the confidence intervals for the mean directions are based on bootstrap calculations. (a) Subsample of size 10. (b) Subsample of size 20. (c) Complete sample.

on the analysis of these data in Fisher & Powell (1989). Following that reference, we also show the various sets of confidence intervals for the three data sets, in Figure 4.12.

It is worth stressing the importance of $\hat{\sigma}$, or equivalently of $\hat{\delta}$. In many applications, particularly in the Earth Sciences, the data set in hand is but

Table 4.1. *Comparison of confidence intervals for median and mean directions, using exact and approximate intervals for the median direction for n = 10 and n ≥ 20 respectively, and bootstrap and large-sample methods for the mean direction.*

Sample size	10	20	30
Median direction $\tilde{\theta}$	279.0°	247.5°	245.0°
Mean direction $\hat{\mu}$	280.8°	248.7°	247.6°
95% Confidence interval for median direction	(245.0°, 315.0°)	(229.0°, 277.0°)	(229.0°, 267.0°)
95% Bootstrap confidence interval for mean direction	(256.6°, 313.5°)	(213.9°, 268.6°)	(225.2°, 262.6°)
95% Large-sample confidence interval for mean direction	—	—	(232.7°, 262.5°)

one of many such sets sampled at different locations throughout a region. The larger-scale statistical problem will be to compare the mean direction of this data set with the mean directions of sets from other locations and, where possible, to form pooled estimates of common mean directions. As we shall see in §**5.3.4** and §**5.3.5**, these comparisons and calculations can be made using just the summary values $n \, (\geq 25)$, $\hat{\mu}$, $\bar{R}$ and $\hat{\sigma}$ for each sample; it is not necessary to have access to the complete individual data sets. (Conversely, if the sample sizes are small, the complete data sets *are* required.)

References and footnotes. The bootstrap confidence intervals are given in Fisher & Hall (1989). The large-sample confidence intervals are due to Fisher & Lewis (1983) and Watson (1983).

4.4.5 Testing a mean direction for a specified value

Suppose we wish to test the hypothesis that $\mu = \mu_0$, against the alternative that $\mu \neq \mu_0$, at some specified level of significance α. If we have already calculated a $100(1 - \alpha)\%$ confidence interval for μ as in §**4.4.4**, we can simply check whether or not μ_0 lies inside that interval: if it does, we accept the hypothesis at the $100\alpha\%$ level, otherwise we reject the hypothesis.

If a suitable confidence interval is not available, we can perform the test directly, the precise nature of the test depending on the sample size n.

(a) $n < 25$. A bootstrap method is required. The general procedure for carrying out this test is described in §**8.4**. We need to decide whether or not to assume that the underlying distribution is symmetric about its mean direction, which can be done using the methods in §**4.4.6**.

(b) $n \geq 25$. Calculate the circular standard error $\hat{\sigma}$ from (4.21), and then the test statistic

$$S = \sin{(\hat{\mu} - \mu_0)}/\hat{\sigma} \tag{4.23}$$

The hypothesis that $\mu = \mu_0$ is rejected if $|S|$ is too large.

Critical values. Compare S with the upper $100(\frac{1}{2}\alpha)\%$ values of the $N(0, 1)$ distribution in Appendix A1.

References and footnotes. The un-calibrated bootstrap test is given in Fisher & Powell (1989). The large-sample test is due to Watson (1983).

4.4.6 Testing for symmetry

Symmetry about an unspecified direction can be investigated informally using the simple plotting procedure described in §4.2. To carry out a formal test, a standard test for linear data can be used. Such a test can be based on the well-known Wilcoxon signed-rank statistic. Calculate the median vector $\tilde{\theta}$ (cf. §2.3.2), and then the values

$$\phi_i = \theta_i - \tilde{\theta}, \quad i = 1,\ldots,n \tag{4.24}$$

Let

$$r_i = \text{rank of } |\phi_i| \text{ amongst } |\phi_1|,\ldots,|\phi_n| \tag{4.25}$$

and calculate

$$w_n^+ = \text{sum of ranks corresponding to positive } \phi_i \text{ values} \tag{4.26}$$

If two or more ϕ_i are tied in absolute value (e.g. the three smallest are equal), assign the average of the corresponding rank values to each (e.g. the three smallest values would get ranks 1, 2, 3, so assign each the rank 2). Data values equal to the median should be deleted and the sample size n correspondingly reduced in the calculations. The null hypothesis of symmetry of the distribution is rejected if w_n^+ is too small or too large.

Critical values.

(a) $n < 16$. For a test of level α, reject the null hypothesis at level α if either

$$w_n^+ \geq w_{n,\frac{1}{2}\alpha} \quad \text{or} \quad w_n^+ \leq n(n+1)/2 - w_{n,\frac{1}{2}\alpha} \tag{4.27}$$

where $w_{n,\frac{1}{2}\alpha}$ is the $\frac{1}{2}\alpha$ critical value for sample size n, given in Appendix A7.

(b) $n \geq 16$. Calculate

$$w^+ = \frac{w_n^+ - n(n+1)/4 + 0.5}{[n(n+1)(2n+1)/24]^{\frac{1}{2}}} \tag{4.28}$$

and hence $\Phi(w^+)$ from Appendix A1. Then the significance probability of the test statistic w is $2[1 - \Phi(w^+)]$.

> **Example 4.17** For the palaeocurrent data of Example 4.3 plotted in Figure 4.3, an informal assessment of the hypothesis of symmetry was made in Example 4.6. For the formal test, we get the median direction as 245°. We remove the two data values actually *equal* to 245°, and subtract 245° from each of the 28 remaining values. The ϕ_i values and their associated ranks

Table 4.2. *Pairs of values* (ϕ_i, r_i) *for carrying out the test of the hypothesis of symmetry in Example 4.17.*

ϕ_i	49	56	84	70	32	36	9	27	−3	−68
r_i	21	22	28	26	17	18	5	14	1	24.5
ϕ_i	12	−68	−16	5	−79	−13	−21	−59	12	22
r_i	6.5	24.5	9.5	3	27	8	11.5	23	6.5	13
ϕ_i	−4	−6	42	−16	45	−31	−30	−21		
r_i	2	4	19	9.5	20	16	15	11.5		

r_i are shown in Table 4.2. Then $w_n^+ = 219$. Using (4.28) with $n = 28$, we get $w^+ = 2.26$. From Appendix A1, the significance probability is approximately $2(1 - 0.988) = 0.024$, so there is some evidence for a departure from the hypothesis of symmetry.

References and footnotes. The Wilcoxon test can be found in most introductory statistical texts. The calculation in (4.28) includes a continuity correction recommended in Lehmann (1975). A specific symmetry test for the circle was proposed by Schach (1969), but its small-sample distribution is unknown.

4.5 Statistical analysis of a random sample of unit vectors from a von Mises distribution

4.5.1 Introduction

The reasons for wanting to fit a particular distribution to data have been canvassed in §**3.1**, as was the issue of why the von Mises distribution can be a natural choice of model in theory and at the same time a difficult model to handle in practice. Assuming that this conflict has been resolved in favour of its use, we are faced with two issues *specific to the use of a parametric model* (in addition to the types of problems considered in §**4.4**), namely,

- *goodness-of-fit*: does the von Mises distribution appear to fit the data?
- *outlier problems*: if there are one or more points suspiciously far from the main data-mass, are they still consistent with the bulk of the data in terms of the sample as a whole being from a von Mises distribution; and how should we cope with their possible undue influence on subsequent statistical analysis of the data?

Assessing adequacy of fit of the von Mises model to data is considered in §**4.5.3**, and the assessment of outliers in §**4.5.4**. §**4.5.5** is concerned with point and interval estimation of the parameters μ and κ of a von Mises distribution, and §**4.5.6** with tests of hypotheses concerning μ and κ.

A final point to emphasise before embarking on the rest of §**4.5** is that we can expect to have difficulties when fitting the von Mises model to dispersed data, except when the sample size is large. 'Dispersed' can be taken to refer to situations in which the underlying distribution has mean resultant length $\rho < 0.7$ (i.e. the von Mises concentration parameter $\kappa < 2.0$); and especially $\rho < 0.45$ ($\kappa < 1.0$).

4.5.2 Test for uniformity against a von Mises alternative

The hypothesis of uniformity corresponds to $\kappa = 0$; the appropriate method is the Rayleigh test described in §**4.3**(*iii*). An interesting use of $\bar{R}$ as an indicator of non-uniformity occurs in a study of methods of monitoring heart-rate variability, by Weinberg & Pfeifer (1984).

4.5.3 Goodness-of-fit for the von Mises model

We describe graphical and formal methods for assessing whether the von Mises model is appropriate for a given sample of data. The possibilities we need to consider are:

- that the model fits, subject possibly to the exclusion of a few outlying points; or
- that the model is inappropriate.

(i) Graphical assessment of goodness-of-fit

The graphical method requires preliminary estimates $\hat{\mu}$ and $\hat{\kappa}$ respectively of the parameters μ and κ, which can be calculated from (2.9) and (4.41). Given a sample of n data points, calculate the quantiles $q_1, \ldots, q_n$ of the corresponding 'best–fitting' von Mises distribution $VM(0, \hat{\kappa})$ as described in §**3.3.6** *Note 1(ii)*. Also calculate

$$z_i = \sin \tfrac{1}{2}(\theta_i - \hat{\mu}) \quad i = 1, \ldots, n \tag{4.29}$$

and rearrange them into ascending order to get $z_{(1)} \leq \cdots \leq z_{(n)}$. Plot

$$(\sin(\tfrac{1}{2}q_1), z_{(1)}), \ldots, (\sin(\tfrac{1}{2}q_n), z_{(n)}) \tag{4.30}$$

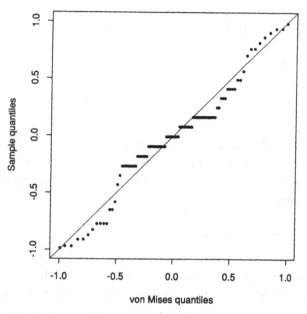

Fig. 4.13 von Mises Q–Q plot for the data in Figure 4.4. Plots of this type can be quite variable, even if the sample has been drawn from the postulated model, so a formal goodness-of-fit test may well be a useful adjunct to the graphical information. See Example 4.18.

If the data have been drawn from a von Mises distribution, the plot should be scattered about a line passing through (0, 0) with a slope of 45°.

Example 4.18 For the data of Example 4.4 (and plotted in Figure 4.4) on the directions chosen by ants in response to a stimulus, suppose we consider attempting to model the data by a von Mises distribution. We get, from (2.9) and (4.41), the parameter estimates $\widehat{\mu} = 183.1°, \widehat{\kappa} = 1.54$. The von Mises Q–Q plot is shown in Figure 4.13. There is a suggestion of systematic departure from the straight line, but these types of plot can be very variable, even when the hypothesised model is appropriate to the data; so it is helpful to supplement the graphical information with a formal test. Such a test for these data is considered below, in Example 4.19.

A useful extra feature of the Q–Q plot is that it may reveal an outlying value, for an otherwise reasonably concentrated sample. Sometimes, outliers will be clearly evident in the raw data display; at other times, they may be more noticeable as extreme points in the plot lying some distance from the 45° line.

Note: with a large sample of very dispersed data, some detail can be lost at the extreme ends of the plot. In such cases, it may be helpful to use instead

$$z_i^* = \sin \tfrac{1}{4}(\theta_i - \hat{\mu}) \qquad (4.31)$$

and plot

$$(\sin(\tfrac{1}{4}q_1), z_{(1)}^*), \ \ldots, \ (\sin(\tfrac{1}{4}q_n), z_n^*) \qquad (4.32)$$

For many data sets, an immediate decision about whether or not the von Mises distribution is a suitable model for the data can be made on the basis of the Q–Q plot. However, this is not always the case, and a supplementary formal statistical test will be desirable.

(ii) Formal test of goodness-of-fit

The following test is suitable for samples of size $n \geq 20$. Suppose first that neither μ nor κ is known. Calculate preliminary estimates of the parameters μ and κ from (2.9) and (4.40), and then the cumulative frequency values (cf. §**3.3.6** *Note 1(ii)*).

$$z_i = F_{\hat{\kappa}}(\theta_i - \hat{\mu}), \ i = 1, \ldots, n \qquad (4.33)$$

Rearrange the z_i into increasing order, to get

$$z_{(1)} \leq \cdots \leq z_{(n)} \qquad (4.34)$$

and then compute the statistic

$$U^2 = \sum_{i=1}^{n} [z_{(i)} - (2i-1)/(2n)]^2 - n(\bar{z} - \tfrac{1}{2})^2 + 1/(12n) \qquad (4.35)$$

where $\bar{z} = \sum_{i=1}^{n} z_{(i)}/n$. The hypothesis that the sample has been drawn from a von Mises distribution is rejected if U^2 is too large.

Critical values. Compare U^2 with the critical values in the appropriate part of the table in Appendix A8.

It may be the case that one or both of the parameters are known, in which event the procedures should be altered as follows:

(a) μ and κ both known. In this case, use $F_\kappa(\theta_i - \mu)$ rather than $F_{\hat{\kappa}}(\theta_i - \hat{\mu})$ in (4.33), calculate U^2 as in (4.35) and then the modified statistic

$$U^* = (U^2 - 0.1/n + 0.1/n^2)(1.0 + 0.8/n) \qquad (4.36)$$

and proceed as above using the appropriate part of Table A8.

(b) μ **known,** κ **unknown.** Estimate κ from (4.40) using $\bar{R}_0 = \sum_{i=1}^{n} \cos(\theta_i - \mu)/n$ instead of $\bar{R}$, and use $F_{\hat{\kappa}}(\theta_i - \mu)$ instead of $F_{\hat{\kappa}}(\theta_i - \hat{\mu})$ in (4.33). Refer U^2 to the appropriate part of Table A8.

(c) κ **known,** μ **unknown.** Estimate μ in the usual fashion, use $F_{\kappa}(\theta_i - \hat{\mu})$ rather than $F_{\hat{\kappa}}(\theta_i - \hat{\mu})$ in (4.33), and refer U^2 to the appropriate part of Table A8.

> **Example 4.19** Continuing the evaluation of the von Mises model (cf. Example 4.18) for the data in Example 4.4, we need the basic estimate $\hat{\kappa}_{ML} = 1.55$ rather than the adjusted estimate used in Example 4.18; although in this case it makes little difference, since $\hat{\kappa}_{ML} = 1.55, \hat{\kappa} = 1.54$. Then the value of the test statistic is $U^2 = 0.32$, which is clearly extreme compared with the values in Table A8, entering the table at $\kappa = 1.5$. We conclude that the von Mises distribution is not a suitable model for the data, confirming the evidence from Example 4.18. Any inference carried out for these data should therefore be done using the large-sample methods described earlier in this chapter.

References and footnotes. The tests are due to Lockhart & Stephens (1985), and are essentially versions of Watson's (1961) statistic customised to von Mises samples with unknown parameter values. The case when both μ and κ are known is just that of Watson's statistic. Other methods recommended in the literature are chi-squared tests (see e.g. Rao & Yoon, 1983), and a score test suggested by Cox (1975) (see also Barndorff–Nielsen & Cox, 1979) and implemented by Lawson (1991); however neither of these approaches appears to perform as effectively as the one recommended here.

4.5.4 Outlier test for discordancy for von Mises data

What is an *outlier* and what is a *discordant observation*? Following Barnett & Lewis (1984), we shall use the word 'outlier' to refer to a surprising observation, that is, to an observation which is *suspiciously far* from the main data mass. The phrase 'discordant observation' will be reserved for a point which is judged to be *significantly far away from the main data mass on the basis of a statistical test*, given some probability model for the way in which the sample arose (e.g. the sample is a random sample from a von Mises distribution). For general information on the subject of outliers, see Barnett & Lewis (1984) and Hawkins (1980); a discussion in the context of spherical data is given in Fisher *et al.* (1987, §5.3.2(iii)). Briefly, some straightforward ways in which outliers can occur are

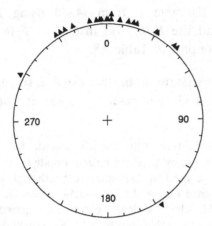

Fig. 4.14 Measurements of resultant directions of 22 Sardinian sea stars after 11 days of movement. See Example 4.20.

(a) from mis-recording of data; or
(b) unwitting sampling from a second population; or
(c) the vagaries of sampling resulting in the occasional isolated value.

As noted elsewhere in this book, outliers in circular data do not usually present serious problems to the data analyst, so we shall dwell on the topic only briefly.

Suppose that one sample value, θ_k say, appears to lie suspiciously far from the main data mass. Typically, the outlier will be detected in a plot of the raw data, or in the von Mises Q–Q plot, although occasionally it will be noted from an inspection of a data listing. We can carry out a formal test of discordancy as follows.

Let R and R_k denote respectively the resultant lengths of the full sample and the sample obtained by deleting θ_k (cf. (2.7)), and calculate the test statistic

$$M_n = \frac{R_k - R + 1}{n - R} \tag{4.37}$$

The outlying point θ_k is classified as discordant if M_n is too large.

(This assumes that the subsample of size $n - 1$ with the largest resultant length will be the one with the outlier θ_k omitted. If this is not the case, there are good grounds for assuming that θ_k is not discordant.)

Critical values. Compare M_n with the values in Appendix A9.

Example 4.20 Figure 4.14 shows a plot of the resultant directions of 22 sea stars 11 days after being displaced from their natural habitat; the data were

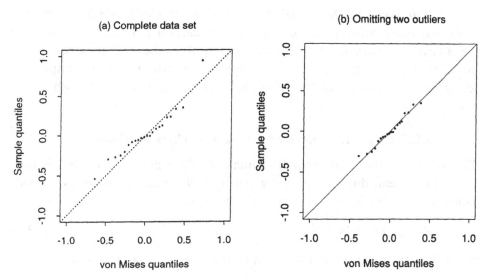

Fig. 4.15 von Mises *Q–Q* plots for the data in Figure 4.14. The effect of the two outliers is clearly evident in (a) and makes assessment of the judgment of fit of the von Mises model difficult. When they are removed, the resulting graph (b) suggests that the model is a reasonable fit to the data. See Example 4.20.

abstracted from a figure in the original paper (Pabst & Vicentini, 1978) by Upton & Fingleton (1989, page 274). (The data are listed in Appendix B11.) The direction of the shore can be taken as approximately 0°. The plot contains two outliers, one sea star which moved along a somewhat lengthier route towards the shore (298°), and another very wayward specimen which navigated out to sea (147°). If we seek to summarise the general characteristics of the data by a von Mises distribution, we might set aside the wayward measurement (possibly an occurrence of case (b) above) and examine the other outlier in line with case (c).

We begin by looking at a probability plot of the whole sample. A von Mises *Q–Q* plot (cf. §**4.5.3**(*i*)) is shown in Figure 4.15(a): the two outliers appear to be having a considerable influence on the graph. If we remove them from the sample and recalculate the *Q–Q* plot, we get the graph in Figure 4.15(b), which suggests that the von Mises model may be a reasonable fit. (A formal test of the type described in §**4.5.3**(*ii*) confirms this.) So, we can consider setting the value 147° aside for the time being and testing the other outlier for discordancy. We get a value of M_n of about 0.30 which, from Table A9 (with a small extrapolation), is not at all extreme. We conclude that the value 298° is consistent with the remaining 20 points, in terms of the von Mises distribution being an adequate model for the 21 points.

References and footnotes. The test is due to Collett (1980), who also evaluated some other possible statistics. In certain circumstances, one of the alternative statistics may perform better than the one described here; however, M_n appears to have good properties overall. A Bayesian approach to the outlier problem has been studied by Bagchi & Guttman (1990).

4.5.5 *Parameter estimation for the von Mises distribution*

The usual (maximum likelihood) estimate $\hat{\mu}$ of the mean direction μ is just the sample mean direction $\hat{\mu} = \bar{\theta}$ given by (2.9). The maximum likelihood estimate $\hat{\kappa}_{ML}$ of κ is the solution of the equation

$$A_1(\hat{\kappa}_{ML}) = R/n = \bar{R} \tag{4.38}$$

where $\bar{R}$ is the mean resultant length of the sample (cf. (2.10)) and

$$A_1(x) = I_1(x)/I_0(x) \tag{4.39}$$

the ratio of two modified Bessel functions. Details about $A_1(x)$ can be found in §**3.3.6**, *Note 1(i)*), and a table of values in Appendix A4. A reasonable approximation to the solution of (4.38) is given by

$$\hat{\kappa}_{ML} = \begin{cases} 2\bar{R} + \bar{R}^3 + 5\bar{R}^5/6 & \bar{R} < 0.53 \\ -0.4 + 1.39\bar{R} + 0.43/(1-\bar{R}) & 0.53 \leq \bar{R} < 0.85 \\ 1/(\bar{R}^3 - 4\bar{R}^2 + 3\bar{R}) & \bar{R} \geq 0.85 \end{cases} \tag{4.40}$$

Unfortunately, $\hat{\kappa}_{ML}$ can be seriously biased when the sample size and $\bar{R}$ are small ($\bar{R} < 0.7$, and particularly $\bar{R} < 0.45$), in that it can substantially over-estimate the true value of κ. For $n \leq 15$, the estimate

$$\hat{\kappa} = \begin{cases} \max\left(\hat{\kappa}_{ML} - 2(n\hat{\kappa}_{ML})^{-1}, 0\right) & \hat{\kappa}_{ML} < 2 \\ (n-1)^3\hat{\kappa}_{ML}/(n^3 + n) & \hat{\kappa}_{ML} \geq 2 \end{cases} \tag{4.41}$$

is to be preferred.

Our choice of method of calculating estimates of the variability of $\hat{\mu}$ and of $\hat{\kappa}$ will depend on both the sample size and on the estimated value of κ.

(i) Confidence interval for the mean direction

Two general approaches are possible.

(a) Parametric bootstrap. Use the general bootstrap method in §**8.3** combined with modified re-sampling (*Technique 3* in *Stage 1*).

Table 4.3. *Guide to selection of estimation or testing method, depending on sample size and size of* $\hat{\kappa}$.

	Method (a)	*Method* (b)
$\hat{\kappa} < 0.4$	all sample sizes	
$0.4 \le \hat{\kappa} < 1.0$	$n < 25$	$n \ge 25$
$1.0 \le \hat{\kappa} < 1.5$	$n < 15$	$n \ge 15$
$1.5 \le \hat{\kappa} < 2.0$	$n < 10$	$n \ge 10$
$\hat{\kappa} \ge 2.0$		all n

(b) Parametric standard error. Estimate the circular standard error of $\hat{\mu}$ by the particular form of the standard error for the von Mises distribution

$$\hat{\sigma}_{VM} = 1/(n\bar{R}\hat{\kappa})^{\frac{1}{2}} \tag{4.42}$$

a $100(1 - \alpha)\%$ confidence interval for μ is then

$$\hat{\mu} \pm \sin^{-1}(z_{\frac{1}{2}\alpha}\hat{\sigma}_{VM}) \tag{4.43}$$

where $z_{\frac{1}{2}\alpha}$ is the upper $100(\frac{1}{2}\alpha)$ percentage point of the Normal $N(0,1)$ distribution in Appendix A1.

The appropriate method can be determined from Table 4.3.

Example 4.21 Figure 4.16 shows a plot of the vanishing directions of 15 homing pigeons, which were released just over 16 kilometres from their loft; the direction of the loft was 149°. (The data are listed in Appendix B12.) Using the goodness-of-fit methods in §**4.5.3** reveals some evidence that a von Mises distribution may not be a totally adequate description for the data (the significance probability of the statistic U^2 in (4.35) lies between 0.1 and 0.05), however we shall take the model as reasonable for the purposes of illustration. The parameter estimates are $\hat{\mu} = 168.5°$, $\hat{\kappa}_{ML} = 2.21$ and $\hat{\kappa} = 1.80$ (based on $\bar{R} = 0.7298$). For purposes of comparison, we shall calculate a confidence interval for μ using both parametric bootstrap methods and the standard error-based approach, since the estimate of κ is on the borderline in terms of choice of methods and the sample size is small. From (4.42), the circular standard error of $\hat{\mu}$ is 0.2252, leading to a 95% confidence interval for $\hat{\mu}$ of $(142.3°, 194.7°)$. (This is fractionally larger than the interval reported by Fisher & Lewis (1983 p. 335), because the more conservative estimate $\hat{\kappa}$ has been used in preference to the maximum likelihood estimate $\hat{\kappa}_{ML}$.)

Using the bootstrap method with $B = 200$ results in a 95% parametric bootstrap confidence interval $\hat{\mu} \pm 24.2° = (144.3°, 192.6°)$. In view of the excellent agreement between these two intervals, the interval based on the circular standard error can be regarded as reliable, and $\hat{\sigma}_{VM}$ could be used

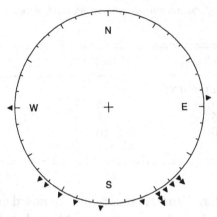

Fig. 4.16 Measurements of the vanishing directions of 15 homing pigeons, released just over 16 kilometres from their loft. (The bearing of the loft was 149°.) See Example 4.21.

in related statistical inference (see e.g. Example 4.23 below). Finally, we present the information in the bootstrap distribution in another way, as suggested in §2.2(*iv*). Using a smoothing parameter of $h = 0.59$, we get the distribution shown in Figure 4.17. Because we have assumed that the underlying distribution is symmetric, it makes sense to build this information into the density estimate.

(ii) Confidence interval for the concentration parameter

Again, we have two possible approaches, either by bootstrap methods or, when $\hat{\kappa}$ is large, by direct calculation.

(a) $\hat{\kappa} < 2$. Use the parametric bootstrap method outlined in §**8.3**, *Stage 1 (Technique 3)*, *Stage 2* and *Stage 3* to obtain, from Algorithm 4, the B bootstrap vectors

$$\begin{pmatrix} C_1 \\ S_1 \end{pmatrix}, \ldots, \begin{pmatrix} C_B \\ S_B \end{pmatrix} \tag{4.44}$$

Then for $i = 1, \ldots, B$, calculate $\bar{R}_i^* = (C_i^2 + S_i^2)^{\frac{1}{2}}$ and finally $\hat{\kappa}_i^*$ using (4.40) and (4.41) with $\bar{R} = \bar{R}_i^*$. To get a $100(1 - \alpha)\%$ confidence interval for κ, sort the bootstrap estimates $\hat{\kappa}_1^*, \ldots, \hat{\kappa}_B^*$ into increasing order:

$$\hat{\kappa}_{(1)}^* \leq \cdots \leq \hat{\kappa}_{(B)}^* \tag{4.45}$$

Set

$$l = \text{integer part of } (\tfrac{1}{2}B\alpha + \tfrac{1}{2}), \quad m = B - l \tag{4.46}$$

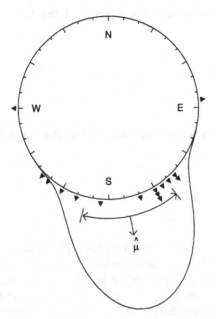

Fig. 4.17 Nonparametric density estimate of the distribution of the sample mean direction, based on 200 parametric bootstrap estimates, for data in Figure 4.16. The sample mean direction and a 95% confidence interval for the true mean are also shown. See Example 4.21.

The desired confidence interval for κ is

$$(\widehat{\kappa}^*_{(l+1)}, \widehat{\kappa}^*_{(m)}) \qquad (4.47)$$

(b) $\widehat{\kappa} \geq 2$. Direct calculation of a confidence interval is possible. Recall that $\chi^2_n(\beta)$ is the upper $100\beta\%$ point of the χ^2_n distribution (Appendix A2) and $\chi^2_n(1 - \beta)$ the lower $100\beta\%$ point, for $0 < \beta < 0.5$.

(b1) A two-sided $100(1 - \alpha)\%$ confidence interval (κ_L, κ_U) Let

$$a = (n - R)/\chi^2_{n-1}(1 - \tfrac{1}{2}\alpha) \qquad (4.48)$$

and

$$b = (n - R)/\chi^2_{n-1}(\tfrac{1}{2}\alpha) \qquad (4.49)$$

Then

$$\kappa_L = [1 + (1 + 3a)^{\frac{1}{2}}]/4a \qquad (4.50)$$

and

$$\kappa_U = [1 + (1 + 3b)^{\frac{1}{2}}]/4b \qquad (4.51)$$

(b2) A one-sided $100(1 - \alpha)\%$ confidence interval of the form $(0, \kappa_U)$

Let

$$b = (n - R)/\chi^2_{n-1}(\alpha) \qquad (4.52)$$

Then

$$\kappa_U = [1 + (1 + 3b)^{\frac{1}{2}}]/4b \qquad (4.53)$$

(b3) A one-sided $100(1 - \alpha)\%$ confidence interval of the form (κ_L, ∞)

Let

$$a = (n - R)/\chi^2_{n-1}(1 - \alpha) \qquad (4.54)$$

Then

$$\kappa_L = [1 + (1 + 3a)^{\frac{1}{2}}]/4a \qquad (4.55)$$

Example 4.22 Continuing the analysis of the homing pigeon data in Example 4.21 (and plotted in Figure 4.16), we shall make a similar comparison of the parametric bootstrap and large-κ parametric confidence intervals for κ, by calculating a two-sided 95% confidence interval using each method. From (4.48) and (4.49), $a = (15 - 10.95)/5.63 = 0.719, b = (15 - 10.95)/26.1 = 0.155$, so that from (4.50) and (4.51), the confidence interval is (1.0, 3.6).

For the bootstrap method, a larger number (2000) of bootstrap samples was used than for estimating the mean direction, because of the asymmetry of the distribution of $\hat{\kappa}$. The resulting 95% confidence interval is (1.4, 11.3).

There is clearly a substantial difference between these two intervals. In view of the small sample size, and small value for $\hat{\kappa}$, the bootstrap interval is probably to be preferred.

References and footnotes. The modified estimate of κ recommended above is due to Best & Fisher (1981). A number of alternative estimates of μ and κ have been proposed. Abeyasekera & Collett (1982) advocated estimates based on the closeness of the von Mises distribution to the Normal distribution on the line, largely on the basis of their simplicity; however, the simplicity is at the cost of other data manipulation beforehand, and the basic estimates recommended above should not present problems with current computing facilities. Lenth (1981) studied robust M-estimates of μ; Ducharme & Milasevic (1987) investigated the sample median direction as an estimate of μ when the sample was contaminated, and discussed circumstances under which it might be preferable to the sample mean direction. Schou (1978) studied an alternative estimate for κ, however its performance for small sample size and small κ is unknown. The confidence interval for the mean direction is based on large-sample theory in Fisher & Lewis (1983) and Watson (1983, Chapter 4). Confidence intervals for μ and κ were implicit in the paper by Watson & Williams (1956) and are given explicitly in Mardia

(1972a) and Cabrera *et al.* (1991) for example. M.A. Stephens developed several approximations providing confidence intervals for small values of κ: see e.g. Stephens (1969a) and references therein, and Mardia (1972a). Upton (1974, 1986) has provided competitive confidence intervals, probably very similar to the one recommended above for large sample size; see also Upton & Fingleton (1989). The view taken here is that, for small values of κ, especially in small samples, the parametric bootstrap is likely to yield more reliable intervals; also, more information is available about the estimate by viewing the whole bootstrap distribution. The bootstrap estimate of κ is due to Fisher & Hall (1991b). With large samples, the approach using the circular standard errors allows the possibility of comparison and combination with other samples in a straightforward fashion (cf. §**5.3.4** and §**5.3.5**). A Bayesian approach to inference about the von Mises distribution has been studied by Mardia & El-Atoum (1976) and Bagchi & Kadane (1991).

4.5.6 Test for a specified mean direction or concentration parameter of a von Mises distribution

(i) Test for specified mean direction

Suppose we wish to test the hypothesis that the mean direction of the distribution from which the sample was drawn is μ_0. In most practical situations, the concentration parameter κ is unknown, and we shall devote most of our discussion to this case. If we have already calculated a $100(1 - \alpha)\%$ confidence interval for the mean direction as in §**4.5.5**, we can carry out a test of level $100\alpha\%$ simply by checking whether this interval covers μ_0. Otherwise, we can proceed directly, with our choice of procedure depending (as it did in §**4.5.4**) on the sample size n, and size of estimate $\hat{\kappa}$. Precisely analogously to §**4.5.5**, the two methods are:

(a) **Parametric bootstrap test.** Use the general bootstrap testing method in §**4.4.5(a)**, except with the modified resampling (*Stage 1, Technique 3*) in §**8.3.2**.

(b) **Parametric test.** Estimate the circular standard error $\hat{\sigma}_{VM}$ of $\hat{\mu}$ as in (4.42), and then calculate the test statistic.

$$E_n = [\sin(\hat{\mu} - \mu_0)]/\hat{\sigma}_{VM} \tag{4.56}$$

Let $z_{\frac{1}{2}\alpha}, z_\alpha$ denote the upper $100(\frac{1}{2}\alpha)\%$ and $100\alpha\%$ points respectively of the Normal $N(0, 1)$ distribution in Table A1. The following tests are then at the $100\alpha\%$ level.

Test of $\mu = \mu_0$ against $\mu \neq \mu_0$: reject if $|E_n| > z_{\frac{1}{2}\alpha}$.

Test of $\mu = \mu_0$ against $\mu < \mu_0$: reject if $\mu_0 - \pi < \widehat{\mu} < \mu_0$ and $E_n < -z_\alpha$.

Test of $\mu = \mu_0$ against $\mu > \mu_0$: reject if $\widehat{\mu} < \mu_0 + \pi$ and $E_n > -z_\alpha$.

The choice of method is determined from Table 4.3.

> **Example 4.23** For the data on vanishing directions of 15 pigeons analysed in Example 4.21, consider testing the null hypothesis that their mean vanishing direction μ is in fact in the direction of their loft (149°), against the alternative that they cannot navigate straight home. From Example 4.21, we have $\widehat{\mu} = 168.5°$ and $\widehat{\sigma}_{VM} = 0.2252$, so that from (4.56), $E_n = 1.48$, which is not at all extreme as a value of a Normal $N(0,1)$ variate. So there is no evidence that the pigeons do not head straight home.

Now suppose that $\kappa = \kappa_0$, a known value.

Again, the size of κ_0 determines the selection of method.

(a*) $\kappa_0 < 2$. Let $\kappa' = \kappa_0 R$, and let $\delta(\kappa', \frac{1}{2}\alpha)$, $\delta(\kappa', \alpha)$ be the upper $100(\frac{1}{2}\alpha)\%$ and $100\alpha\%$ points respectively of the von Mises $VM(0, \kappa')$ distribution, which can be calculated using the algorithm in §**3.3.6**, *Note 1(ii)*. The following tests are then at the $100\alpha\%$ level.

Test of $\mu = \mu_0$ against $\mu \neq \mu_0$: reject if $|\widehat{\mu} - \mu_0| > \pi - \delta(\kappa', \frac{1}{2}\alpha)$.

Test of $\mu = \mu_0$ against $\mu < \mu_0$: reject if $\mu_0 - \widehat{\mu} > \pi - \delta(\kappa', \alpha)$.

Test of $\mu = \mu_0$ against $\mu > \mu_0$: reject if $\widehat{\mu} - \mu_0 > \pi - \delta(\kappa', \alpha)$.

(b*) $\kappa_0 \geq 2$. Proceed as in the case κ unknown but $\widehat{\kappa} \geq 2$, except that $\widehat{\sigma}_{VM}$ is replaced by

$$\sigma_{VM} = 1/(n\rho\kappa_0)^{\frac{1}{2}} \tag{4.57}$$

in (4.56), where ρ can be calculated from κ_0 using Appendix A3 (cf. §**3.3.6** *Note 1(i)*).

(ii) Test for a specified concentration parameter

As with testing a hypothesis about the mean direction (§**4.5.6(i)**), we can carry out a test immediately if we have already calculated an appropriate confidence interval for κ. Alternatively, we can perform a test directly, the precise form of the test depending on the *estimated value* $\widehat{\kappa}$. There are slight modifications in the rarely encountered situation in which μ is known. Suppose first that μ is unknown.

(a) $\hat{\kappa} < 2$. Use a bootstrap method. To avoid a proliferation of descriptions of customised bootstrap methods, we recommend that the testing approach via confidence intervals be used. By looking at the complete set of bootstrap values, a confidence interval can be found which *just* covers the hypothesised value κ_0. If this is a $100(1 - \beta)\%$ interval, the *significance probability* for the associated test is β.

(b) $\hat{\kappa} \geq 2$. (See §4.5.5(*ii*)(*b*) to confirm notation.) Let $\gamma_0 = 1/[1/\kappa_0 + 3/(8\kappa_0^2)]$.

The following are tests of level $100\alpha\%$. Test of $\kappa = \kappa_0$ against $\kappa \neq \kappa_0$: reject if either

$$R < n - \chi_{n-1}^2(\tfrac{1}{2}\alpha)/2\gamma_0$$

or

$$R > n - \chi_{n-1}^2(1 - \tfrac{1}{2}\alpha)/2\gamma_0$$

Test of $\kappa = \kappa_0$ against $\kappa < \kappa_0$: reject if

$$R < n - \chi_{n-1}^2(\alpha)/2\gamma_0$$

Test of $\kappa = \kappa_0$ against $\kappa > \kappa_0$: reject if

$$R > n - \chi_{n-1}^2(1 - \alpha)/2\gamma_0$$

Now suppose that $\mu = \mu_0$, a known direction.

(a*) $\hat{\kappa} < 2$. Use a bootstrap method and a confidence interval approach, as recommended in *(a)* above. With $\mu = \mu_0$ known, μ_0 can be subtracted from each sample point to centre the sample at 0, so we assume this has been done, giving the values $\psi_1, \ldots, \psi_n$. The resultant length of the sample is now calculated from $\psi_1, \ldots, \psi_n$ as in (2.7), and κ estimated from this (4.41). Resamples should be centred at $\hat{\mu}$, the mean of $\psi_1, \ldots, \psi_n$, and bootstrap estimates of κ estimated from R_0^*. Calculate a confidence interval for κ as in §4.5.5(*ii*)(*a*). For small sample sizes, it will probably be necessary to calibrate the confidence interval using the method in §8.4.3. Then proceed to testing, as described in *(a)*.

(b*) $\hat{\kappa} \geq 2$. This is essentially the same as case *(b)*, immediately above, except that R is replaced by $R_0 = \sum_{i=1}^n \cos(\theta_i - \mu_0)$, and the χ_{n-1}^2 percentage points are replaced by $\chi_n^2(\tfrac{1}{2}\alpha)$, $\chi_n^2(1 - \tfrac{1}{2}\alpha)$, $\chi_n^2(\alpha)$ and $\chi_n^2(1 - \alpha)$ as appropriate.

Example 4.24 Continuing the analysis of Example 4.22, suppose we wish to test the hypothesis that $\kappa = 1.5$, against the alternative that $\kappa < 1.5$. Since $\hat{\kappa} < 2$ we choose a bootstrap method. A one-sided confidence interval of the form (κ_L, ∞) is needed. From the 2000 bootstrap values for κ in Example 4.22, we get a one-sided 95% interval to be $(1.48, \infty)$, which covers the hypothesised value. Alternatively, a confidence interval of this form which *just* covers $\kappa = 1.5$ starts from the 116th value in the ordered sequence of 2000 bootstrap estimates, so the significance probability for the test is about 0.057. Thus there is some evidence that κ is larger than 1.5.

References and footnotes. The tests for large values of κ are due to Stephens (1969a); some improvements on these were obtained by Upton (1973). For other relevant references and comments, see the corresponding notes to §**4.5.5**.

4.6 Statistical analysis of a random sample of unit vectors from a multimodal distribution

Samples of multimodal data arise in a variety of settings, such as those in Examples 2.1, 2.2 and 2.9. In geological applications, typical questions of interest are:

- How many modal groups (unimodal populations) are present in the sample?

- What are the mean directions of these groups?

- What are the relative proportions of data in each group?

In Example 2.10, nonparametric density estimation was used to get a partial answer to the first of these questions. Given that answer, *ad hoc* methods can then provide approximate answers to the other questions. However, it is difficult to proceed further without making more statistical assumptions, because no effective formal statistical test is available at present for testing hypotheses about the number of modes.

If one is prepared to assume that the sample has been drawn from a mixture of different von Mises distribution, *with the number of component modal groups known*, then progress can be made on the other questions. In §**4.6.1**, we consider the case of two components, but the method extends more generally – if rather more expensively – to more than two. A test for determining the number of (von Mises) components in a mixture is given in §**4.6.2**.

It should be remarked, however, that attempts to get good estimates of the parameters of a mixture of distributions will generally be in vain, unless either the raw data plot is clearly bimodal (or multimodal) or the sample size is large.

4.6.1 Fitting a mixture of two von Mises distributions

Suppose that the sample $\theta_1, \ldots, \theta_n$ has been drawn from a distribution comprising an unknown proportion p from a $VM(\mu_1, \kappa_1)$ population and the remaining unknown proportion $1 - p$ from a $VM(\mu_2, \kappa_2)$ population. The objective is to estimate the five unknown parameters $\mu_1, \mu_2, \kappa_1, \kappa_2$ and p. Set up the following six equations:

$$\left. \begin{array}{l} pA_1(\kappa_1)\cos(\mu_1) + (1-p)A_1(\kappa_1)\cos(\mu_2) = \bar{C}_1 \\ pA_2(\kappa_1)\cos(2\mu_1) + (1-p)A_2(\kappa_1)\cos(2\mu_2) = \bar{C}_2 \\ pA_3(\kappa_1)\cos(3\mu_1) + (1-p)A_3(\kappa_1)\cos(3\mu_2) = \bar{C}_3 \\ pA_1(\kappa_1)\sin(\mu_1) + (1-p)A_1(\kappa_1)\sin(\mu_2) = \bar{S}_1 \\ pA_2(\kappa_1)\sin(2\mu_1) + (1-p)A_2(\kappa_1)\sin(2\mu_2) = \bar{S}_2 \\ pA_3(\kappa_1)\sin(3\mu_1) + (1-p)A_3(\kappa_1)\sin(3\mu_2) = \bar{S}_3 \end{array} \right\} \tag{4.58}$$

where

$$\bar{C}_r = \sum_{i=1}^{n} \cos(r\theta_i)/n, \quad \bar{S}_r = \sum_{i=1}^{n} \sin(r\theta_i)/n \tag{4.59}$$

for $r = 1, 2, 3$ (cf. (2.14), (2.15)) and A_1, A_2 and A_3 are defined in (3.36). These equations have to be solved iteratively to get the five *method of moments* estimates $\hat{\mu}_{mm_1}, \hat{\mu}_{mm_2}, \hat{\kappa}_{mm_1}, \hat{\kappa}_{mm_2}$ and $\hat{p}$. Let $\tilde{\mu}_1, \tilde{\mu}_2, \tilde{\kappa}_1, \tilde{\kappa}_2$ and $\tilde{p}$ be any set of estimates of the parameters, and define the residual differences between the sample moments $(\bar{C}_1, \bar{C}_2, \bar{C}_3, \bar{S}_1, \bar{S}_2, \bar{S}_3)$ and corresponding fitted values (left hand sides of (4.58)) evaluated at the estimates $\tilde{\mu}_1, \tilde{\mu}_2, \tilde{\kappa}_1, \tilde{\kappa}_2$ and $\tilde{p}$) by $\Delta C_1, \Delta C_2, \Delta C_3, \Delta S_1, \Delta S_2$ and ΔS_3; e.g.

$$\Delta C_1 = \tilde{p}A_1(\tilde{\kappa}_1)\cos(\tilde{\mu}_1) + (1-\tilde{p})A_1(\tilde{\kappa}_1)\cos(\tilde{\mu}_2) - \bar{C}_1$$

Define the least squares criterion

$$r^2(\tilde{\mu}_1, \tilde{\mu}_2, \tilde{\kappa}_1, \tilde{\kappa}_2, \tilde{p}) = \Delta C_1^2 + \Delta C_2^2 + \Delta C3^2 + \Delta S_1^2 + \Delta S_2^2 + \Delta S_3^2 \tag{4.60}$$

Then the desired estimates $\hat{\mu}_{mm_1}, \hat{\mu}_{mm_2}, \hat{\kappa}_{mm_1}, \hat{\kappa}_{mm_2}$ and $\hat{p}$ can be found by minimising the criterion $r^2(\tilde{\mu}_1, \tilde{\mu}_2, \tilde{\kappa}_1, \tilde{\kappa}_2, \tilde{p})$ as a function of $\tilde{\mu}_1, \tilde{\mu}_2, \tilde{\kappa}_1, \tilde{\kappa}_2$ and $\tilde{p}$. Algorithms to do this can be found in the NAG library (Algorithm E04JAF, NAG 1991) or IMSL (Algorithm UNLSF, IMSL, 1991), for example. (To get starting values, one can fit the three-parameter mixture described below, which requires no iteration.)

Approximate confidence intervals for μ_1 and μ_2 can be obtained as follows. Set n_1 = integer part of $\hat{p}n+0.5$, $n_2 = n-n_1$, and calculate adjusted estimates $\hat{\kappa}_1$ and $\hat{\kappa}_2$, say, using (4.41). (So, for example, substitute $\hat{\kappa}_1$ for $\hat{\kappa}_{ML}$ and n_1 for n in (4.41) to get the adjusted estimate $\hat{\kappa}_1$; and similarly for $\hat{\kappa}_2$.) Then obtain

$$\hat{p}_1 = A_1(\hat{\kappa}_1), \quad \hat{p}_2 = A_1(\hat{\kappa}_2) \tag{4.61}$$

from (3.36), (3.43) and (3.44). Use n_1, $\hat{p}_1$ and $\hat{\kappa}_1$ to calculate the circular standard error $\hat{\sigma}_{VM1}$ say, from (4.42); similarly compute $\hat{\sigma}_{VM2}$. Finally calculate approximate confidence intervals from (4.43).

Notice that this approach to estimating the parameters of the components of the mixture assumes that the mean directions μ_1 and μ_2 differ, and similarly that the concentrations κ_1 and κ_2 differ. We comment on this further, shortly.

> **Example 4.25** Consider the data displayed in Figure 2.10 and studied in Example 2.9, on the movements of 76 turtles after experimental treatment. It seems reasonable to assume that there are two modal groups present. Following the analysis of these data by Spurr & Koutbeiy (1991), who proposed the estimation method described above, the parameter estimates are found to be
>
> $$\hat{\mu}_{mm_1} = 63.2°, \quad \hat{\mu}_{mm_2} = 240.2°, \quad \hat{\kappa}_{mm_1} = 2.91,$$
>
> $$\hat{\kappa}_{mm_2} = 4.81, \quad \hat{p} = 0.82$$
>
> whence
>
> $$n_1 = 62, \quad \hat{\kappa}_1 = 2.77, \quad \hat{\sigma}_{VM1} = 0.0858,$$
>
> $$n_2 = 14, \quad \hat{\kappa}_2 = 3.83, \quad \hat{\sigma}_{VM2} = 0.1475$$
>
> and the approximate confidence intervals for μ_1 and μ_2 are $63.2° \pm 9.7° = (53.5°, 72.9°)$ and $240.2° \pm 16.8° = (223.4, 257.0)$ respectively.
>
> We might wonder whether the two groups of turtle movements are, in fact, drawn from the same distribution apart from being diametrically opposed, since the difference between the sample estimates is close to 180°, and the dispersions are not vastly different. We can carry out a quick check on this by subtracting 180° from data points lying in the arc $(180°, 360°)$, then doubling all the data (see Figure 4.18) and carrying out goodness-of-fit tests for the von Mises model. This is not a particularly sensitive approach (because we have specific alternative models in mind but we are not using this extra information in the tests) but can produce rapid feedback. A von Mises Q–Q plot (cf. §4.5.3(*i*)) for the altered data set is shown in Figure 4.19. There is little evidence here that the von Mises model is not

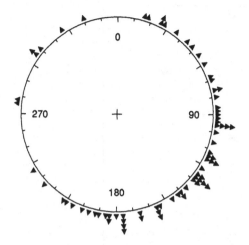

Fig. 4.18 Data derived from Figure 2.10 by reversing directions between 180° and 360°, and doubling all measurements to convert them to vectors. There is little evidence of bimodality, suggesting that the mean direction of both groups of turtles is along a common axis. See Example 4.25.

fitting satisfactorily. The goodness-of-fit statistic U^2 (see (4.35)) has the value 0.024, again providing no evidence against the hypothesis that the mean directions of the two groups lie along a common axis.

Some special cases of this mixture model can arise, as already suggested in the Example:

(a) angle between μ_1 and μ_2 known (e.g. 180°)
(b) dispersions assumed equal
(c) dispersions assumed equal and the objective is to estimate the angle between μ_1 and μ_2

A particular version combining (a) and (b) is the so-called three-parameter mixture: the mean directions are μ and $\mu + 180°$, both dispersions equal κ, and the mixing proportion p is unknown. The parameters of this model can be estimated as follows. Calculate

$$\psi_i = \theta_i[\text{mod } 180°], \ i = 1, \ldots, n \tag{4.62}$$

$$C_\psi = \sum_{i=1}^{n} \cos(2\psi_i), \ S_\psi = \sum_{i=1}^{n} \sin(2\psi_i) \tag{4.63}$$

$$\bar{R}_\psi = (C_\psi^2 + S_\psi^2)^{\frac{1}{2}} \tag{4.64}$$

and hence $\hat{\mu}_\psi$ say from (2.9). Then the parameter estimates can be found

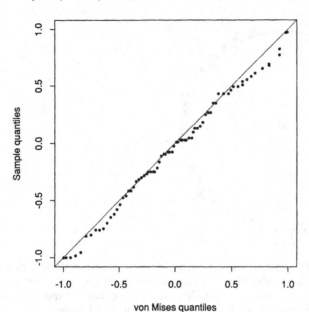

Fig. 4.19 von Mises Q–Q plot for the data in Figure 4.18. There is little indication that the von Mises distribution is not an adequate model for the data. See Example 4.25.

from

$$\widehat{\mu} = \tfrac{1}{2}\widehat{\mu}_\psi \tag{4.65}$$

$$\bar{R}_\psi = A_2(\widehat{\kappa}_{mm}) \tag{4.66}$$

$$(2\widehat{p} - 1)A_1(\widehat{\kappa}_{mm}) = \bar{C}_1 \cos(\widehat{\mu}) + \bar{S}_1 \sin(\widehat{\mu}) \tag{4.67}$$

where $\bar{C}_1$ and $\bar{S}_1$ are defined in (4.59) and $A_2(x)$ is defined in (3.36). The 'method-of-moments' estimate $\widehat{\kappa}_{mm}$ can then be adjusted to $\widehat{\kappa}$ using (4.41). The circular standard error of $\widehat{\mu}$ is then $1/[n\widehat{\kappa}A_1(\widehat{\kappa})]^{\frac{1}{2}}$ (where $A_1(x)$ is defined by (3.36)) and a confidence interval for μ can then be constructed from (4.43).

The special case described in *(c)* has been studied in detail by Bartels (1984).

4.6.2 *A test for the number of components in a mixture of von Mises distributions*

It is possible, if computationally expensive, to test for the number of components in a von Mises mixture by a stepwise procedure (that is, we test for one component against more than one; if there is strong evidence against the hypothesis of one component, we test for two components against more than two, and so on). Here, we sketch the method given by Hsu *et al.* (1986).

For any positive integer r, let

$$f_r(\theta) = \sum_{j=1}^{r} p_j f(\theta; \mu_j, \kappa_j) \tag{4.68}$$

where $f(\theta; \mu_j, \kappa_j)$ is the probability density function (pdf – see §3.2.1) of a von Mises $VM(\mu_j, \kappa_j)$ distribution and p_j is the proportion of this density represented in the overall mixture distribution, $j = 1, \ldots, r$. The *cumulative distribution function* (or cdf – see §3.2.1) corresponding to $f_r(\theta)$ of a mixture of von Mises $VM(\mu_j, \kappa_j)$ distributions with weights p_j is then given by

$$F_r(\theta) = \sum_{j=1}^{r} p_j F(\theta; \mu_j, \kappa_j) \tag{4.69}$$

where $F(\theta; \mu_j, \kappa_j)$ is the cdf corresponding to $f(\theta; \mu_j, \kappa_j)$.

Our object is to find the value of r for the data set in hand, assuming that it can be modelled by a finite mixture of von Mises distributions. At step k, $k = 1, 2, \ldots$, carry out the following test:

> ***Step 1.*** Estimate the parameters of the components of a mixture of k von Mises distributions and hence the estimated pdf $\widehat{f}_k(\theta)$ and estimated cdf $\widehat{F}_k(\theta)$ by substituting fitted densities and parameter estimates $\widehat{p}_j$ in (4.69). At step $k = 1$, use §4.5.5; for $k = 2$, use the method of §4.6.1; for $k > 2$, use an extended version of this method, involving the estimation of $k - 1$ independent proportions $p_1, \ldots, p_{k-1}$, $p_1 + \cdots + p_k = 1$.
>
> ***Step 2.*** Calculate a value of a goodness-of-fit statistic T for testing the fit of the estimated distribution to the sample $\theta_1, \ldots, \theta_n$. Here, for example, we calculate the statistic $T = U^2$ in (4.35), using $\widehat{F}_k(\theta)$ instead of F as used there.
>
> ***Step 3.*** Simulate a parametric resample $\theta_1^*, \ldots, \theta_n^*$ from $\widehat{F}_k(\theta)$ as follows:
>
> **3.1** Define
> $$\widehat{P}_0 = 0, \ \widehat{P}_j = \widehat{p}_1 + \cdots + \widehat{p}_j, \ j = 1, \ldots, k, \ \widehat{P}_k = 1 \tag{4.70}$$
>
> **3.2** Simulate $u_1, \ldots, u_n$ from the Uniform distribution on (0,1)
>
> **3.3** If $\widehat{P}_{j-1} \le u_i < \widehat{P}_j$, simulate θ_i^* from $VM(\widehat{\mu}_j, \widehat{\kappa}_j)$

Step 4. Estimate the parameters of the components of a mixture of k von Mises distributions for this bootstrap sample, calculate the fitted cdf $\widehat{F}_k^*(\theta)$, and then the goodness-of-fit statistic U^2 to test the fit of $\widehat{F}_k^*(\theta)$ to $\theta_1^*, \ldots, \theta_n^*$, giving a value T_1^*, say.

Step 5. Repeat *Step 3* to obtain a total of B bootstrap samples and test values $T_1^*, \ldots, T_B^*$.

Step 6. The significance probability of the test for k modes against more than k modes is given by

$$(1/B) \sum_{b=1}^{B} I[T \le T_b^*] \tag{4.71}$$

where $I[T \le T_b^*] = 1$ if $T \le T_b^*$, else 0.

Clearly this procedure will require a lot of computing time, since at each bootstrap step a complete function minimisation has to be performed. However, parameter estimates obtained from the initial model fit should provide good starting values in fitting the bootstrap models. The problem of testing nested hypotheses also arises, and can be countered somewhat by accepting lower than usual values of the significance probabilities. Also, in practice, one would start with a sensible minimum value for k – e.g. it is often the case that at least two modes are evident in the data, so the first test would be for two modes against three or more.

References and footnotes. Jones & James (1969) studied a maximum likelihood approach to this problem; other writers investigated alternative methods including the method of moments. Mardia & Sutton (1975) gave a table indicating conditions under which a mixture of two von Mises distributions would have one mode or two modes. Spurr & Koutbeiy (1991) recommended the procedure given here which, despite having more estimating equations than unknown parameters, was found to work better than systems involving just five variables and to give answers comparable to those arising from maximum likelihood. Some special cases of the mixture problem were studied by Stephens (1969b), Mardia (1972a) and Bartels (1984), as noted above.

4.7 Other topics

The foregoing material does not cover the complete range of methods available for analysis of a single sample of data. There are a number of other developments which may be of value in some applications.

(i) Fitting the Wrapped Cauchy distribution. This model was described in §**3.3.4**. Methods for estimating the parameters of this distribution have been suggested by Kent & Tyler (1988).

(ii) Sequential analysis. With sequential analysis, statistical decisions about a population characteristic are made on an ongoing basis as data are collected, in contrast to the methods described above, which are applicable when the whole data set is available for analysis. Thus, data acquisition stops when, for example, a mean direction is known to specified precision or a decision about a null hypothesis can be made at a certain level of significance. Aspects of this approach have been studied by Gadsden & Kanji (1981, 1982).

5

Analysis of two or more samples, and of other experimental layouts

5.1 Introduction

In many situations, the data set under consideration arises not as a single sample of measurements of a single phenomenon, but in the form of samples measured under a variety of experimental conditions or collected from a variety of localities. For such data, interest usually centres on two issues: first, whether there appear to be real differences between the various responses (usually, the mean directions of the samples), if so which are the responses which differ and whether there is sound evidence that the differences really exist (in other words that they are *statistically significant*); second, for those responses which are regarded as comparable, how to combine them to get a pooled estimate of their common mean direction. For the most part, it will be assumed that the individual samples have been drawn from unimodal distributions. The sorts of applications we have in mind are the following:

Example 5.1 Figure 5.1 shows plots of three samples of cross–bed azimuths from the Bulgoo Formation, Belford Anticline. (The data are listed in Appendix B6; Set 2 was studied in Chapter 4, Examples 4.3, 4.13, 4.14, 4.16 and 4.17.) It is of interest to decide whether the samples have been drawn from populations with a common mean direction, and if so, to form a pooled estimate of this common mean direction.

Example 5.2 Figure 5.2 shows samples of orientations of termite mounds of *Amitermes laurensis*, and their mean orientations, at 14 sites in Cape York Peninsula, North Queensland. (The data are listed in Appendix B13.) It is of interest to determine whether the mean orientations are consistent; a related question is whether these mean orientations are consistent with an axis passing through magnetic North.

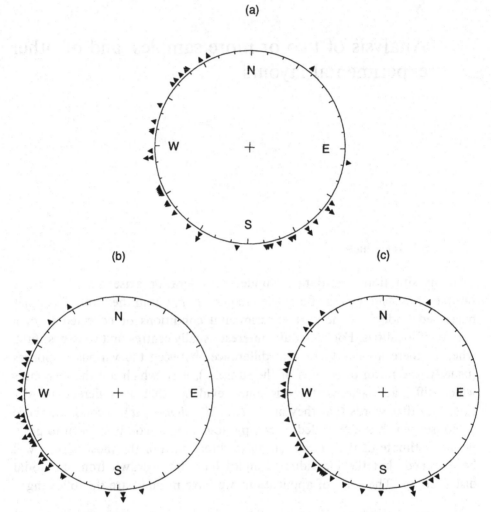

Fig. 5.1 Three samples of measurements of cross-bed palaeocurrents in the Belford Anticline, New South Wales. (a) 40 measurements (b) 30 measurements (c) 30 measurements. See Example 5.1.

Example 5.3 Figure 5.3 shows plots of the wind direction at Gorleston, England between 11.00am and noon on Sundays during 1968, recorded to the nearest 10° and grouped by season. (The data are listed in Appendix B14.) One question of interest is whether some of the seasons exhibit the same distributional pattern.

Example 5.4 Figure 5.4 shows plots of groove and tool marks (axial data) and flute marks (vectorial data) measured in the Murruin Creek area, southwest of Yerranderie, New South Wales. (The data are listed

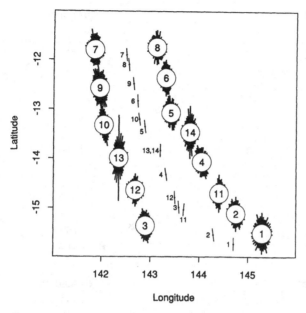

Fig. 5.2 Locations, mean orientations and data plots of samples of orientations of termite mounds in Cape York Peninsula, North Queensland. It is of interest to determine whether the mean orientations are consistent (and possibly, consistent with their being oriented North–South). See Example 5.2.

in Appendix B15.) It is of interest to determine whether the mean trend or orientation of the grooves is consistent with the mean direction of the flutes.

Some exploratory analysis is usually carried out on individual samples (cf. §4.2) to gain an appreciation of the general features of each sample. In particular, we would exclude from further analysis any data set deemed to be from a uniform distribution (cf. §4.3). Then, depending on the amount of common structure we are prepared to assume about the underlying distributions (e.g. that they are all symmetric, or that they are identical except possibly for rotation differences) we can select a method which makes best use of this assumed structure. Thus, the techniques range from simple comparison of median directions, assuming only that each distribution is unimodal, to comparison of mean directions of von Mises distributions.

We begin, in Section 5.2, with a brief discussion of exploratory techniques useful in comparing samples of unit vectors. In Section 5.3, general methods are described for comparing the *reference* directions (mean or median

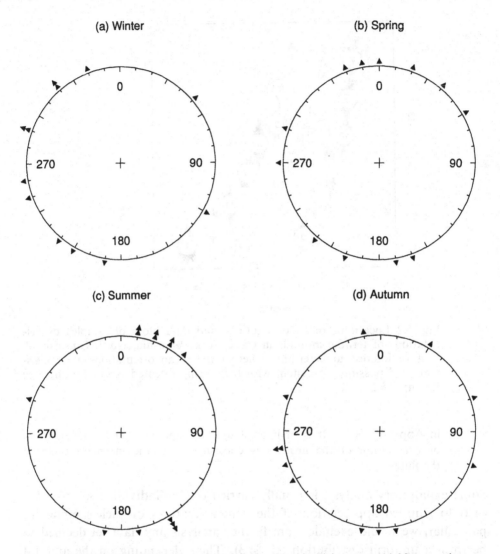

Fig. 5.3 Wind direction at Gorleston, England between 11.00am and noon on Sundays during 1968, grouped by season. (a) Winter (b) Spring (c) Summer (d) Autumn. See Example 5.3.

directions) of two or more samples of unit vectors and, where appropriate, of combining the information from the separate samples to get a *pooled estimate* of a common reference direction and an assessment of the error of this estimate. The same questions are treated in Section 5.4, with the added assumption that the data can be modelled adequately by von Mises distributions. Assuming this, it is also possible to compare and, where ap-

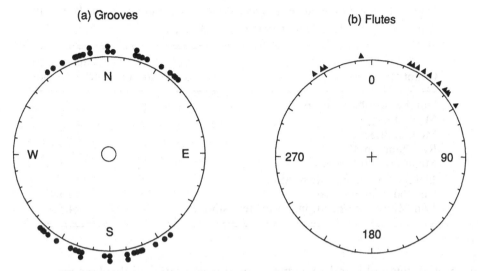

Fig. 5.4 (a) Groove and tool marks (axial data), and (b) flute marks (vectorial data) measured in the Murruin Creek area, southwest of Yerranderie, New South Wales. See Example 5.4.

propriate, combine the estimates of the concentration parameters of the von Mises distributions.

Experimental designs can, of course, be a lot more complex than the so-called one-way layouts, or several-sample problems, described above. In Section 5.5, we make some general remarks about how to approach the analysis of data arising from more complicated experimental layouts.

Because of the heavy emphasis on several-sample problems, it will be useful to standardise the notation for the discussion. Generally, we suppose that there are r samples ($r = 2, 3, \ldots$) with the data in the ith sample being

$$\theta_{i1}, \ldots, \theta_{in_i}, \quad i = 1, \ldots, r \tag{5.1}$$

Notation for the summary statistics calculated from each sample is shown in Table 5.1. The total number of data points is

$$N = n_1 + \cdots + n_r \tag{5.2}$$

5.2 Exploratory analysis

As was the case with the analysis of a single sample of data, choice of statistical method should depend on the assumptions which can reasonably be made. To a large extent, we can determine this choice by analysing the

Table 5.1. *Notation used for summary statistics calculated from the n_i unit directions $\theta_{i1} \ldots, \theta_{in_i}$ in the ith sample, $i = 1, \ldots, r$.*

Quantity	Notation	Reference to earlier section
Number of observations	n_i	
Mean direction	$\bar{\theta}_i$	§2.3.1
Median direction	$\tilde{\theta}_i$	§2.3.2
Resultant length	R_i	§2.3.1
Mean resultant length	$\bar{R}_i$	§2.3.1
Estimate of circular dispersion	$\hat{\delta}_i$	§2.3.1
Circular standard error	$\hat{\sigma}_i$	§4.4.4
Von Mises concentration parameter estimate	$\hat{\kappa}_i$	§4.5.5

samples individually, and by some comparative graphical displays which we now describe.

In §2.2, several displays for a single data set were illustrated. Any of them can be used to show several data sets simultaneously. In particular, several raw data plots, each with an associated nonparametric density estimate (§2.2(*iv*)), can often highlight key differences between the data sets; this is particularly so for multimodal data sets. One display not mentioned in Chapter 2 was the *boxplot*, which is a method used for displaying a sample of linear data so as to emphasise the key distributional features. The boxplot is not helpful for samples of circular data in which the data are multimodal or distributed right around the circle. However, it can be helpful to compare unimodal samples. Details of its construction can be found, for example, in Tukey (1977) or Cleveland (1993). Its use is readily appreciated by studying an example.

> **Example 5.5** Boxplots for the termitaria described in Example 5.2 are shown in Figure 5.5. The ends of a box correspond to the quartiles of the sample, so that the height of the box is the interquartile range (*IQR*). The line across the box corresponds to the median direction (or in this case, axis), or 50% point of the sample. The dashed lines, or *whiskers* extend to the last data value on each side of the median which is not more than $1.5 \times IQR$ from the box; any data lying outside this span are plotted as individual values. The interpretation to be placed on the whiskers is that, for samples from von Mises distributions, only a small percentage of the data values – at most 1–2%, depending on κ – should lie outside the whiskers. We see in this plot that the samples appear to have comparable mean orientations, but that their dispersions are rather different. Most of the sample distributions seem to be reasonably symmetric, judging by the facts that their medians

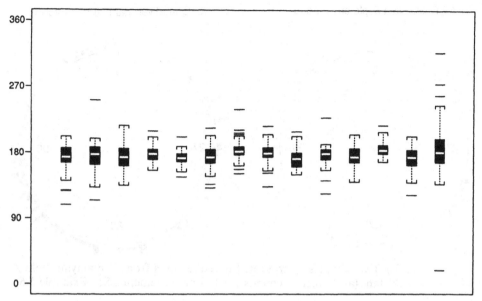

Fig. 5.5 Boxplots for the orientations of termitaria in Figure 5.2, in order of decreasing latitude. See Example 5.5.

plot near the centres of the boxes, and the whiskers on each side appear to be of similar lengths.

A graph which can be used to compare the general distributional shapes of two samples is the angular Q–Q plot. It will *not* pick up differences in reference direction, because the method relies on each sample being centred at its median direction; however such differences are usually spotted quite readily. In the notation introduced in Table 5.1, calculate the two sets of numbers

$$z_{1j} = \sin(\tfrac{1}{2}(\theta_{1j} - \tilde{\theta}_1)), \quad j = 1, \ldots, n_1 \tag{5.3}$$

and

$$z_{2j} = \sin(\tfrac{1}{2}(\theta_{2j} - \tilde{\theta}_2)), \quad j = 1, \ldots, n_2 \tag{5.4}$$

and re-order each set to obtain the sequences

$$z_{(11)} \le z_{(12)} \le \cdots \le z_{(1n_1)} \tag{5.5}$$

and

$$z_{(21)} \le z_{(22)} \le \cdots \le z_{(2n_2)} \tag{5.6}$$

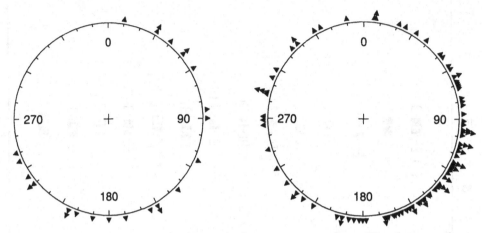

Fig. 5.6 Two samples of cross-bed measurements from Himalayan molasse in Pakistan. (a) 25 measurements (b) 104 measurements. See Example 5.6.

Suppose that $n_1 \leq n_2$, and define the integers

$$j_k = \text{integer part of } n_2 \frac{(k - \frac{1}{2})}{n_1} + 1, \quad k = 1,\ldots,n_1 \tag{5.7}$$

Then plot the pairs of points

$$(z_{(1k)}, z_{(2j_k)}), \quad k = 1,\ldots,n_1 \tag{5.8}$$

If the distributions have similar shape, the points should plot approximately along a 45° line. If one distribution is more dispersed than the other, the points will snake around the line, below at one end and above at the other.

> **Example 5.6** Figure 5.6 shows two samples of cross-bed measurements collected from Himalayan molasse in Pakistan. (The data are listed in Appendix B16.) We shall calculate a Q–Q plot to get an informal idea of the similarity of their distributional shapes. The gap in the data values in the ripple cross-beds is not matched in the other sample; one might hypothesise that this could be explained by the much greater sample size. The Q–Q plot in Figure 5.7 dispels this notion, as it lies almost completely above the 45° line.

References and footnotes. Several boxplot-based displays for representing wind speed and direction at different times during the day were suggested by Graedel (1977).

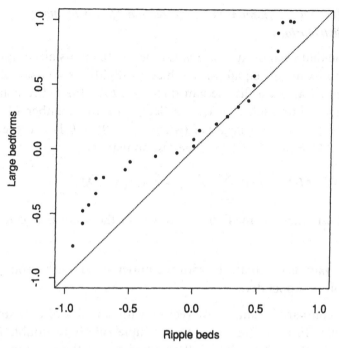

Fig. 5.7 *Q–Q* plot to compare distributional shapes of cross-bed samples in Figure 5.6. See Example 5.6.

5.3 Nonparametric methods for analysing two or more samples of unimodal data

5.3.1 Introduction

This subsection uses the same sorts of methods as were described in §**4.4** for analysing a single sample of unimodal vector data. Our main interest is in comparing the 'preferred' directions (either median or mean directions) and, if possible, in combining the separate sample estimates to form a pooled estimate of a common reference direction. The remarks in §**4.4.1** about the relative merits of the sample median direction and the sample mean direction apply equally here. Accordingly, we indicate the basic uses of median-based methods, but spend rather longer on mean-based methods. In particular, the importance of being able to summarise a single sample with a number of mean-based statistics (cf. the remarks at the end of §**4.4.4**) from the point of view of synthesising information from several sources cannot be over-emphasised.

Occasionally, one wishes to test for complete equality of two or more distributions. A test for this purpose is given in §**5.3.6**.

5.3.2 Test for a common median direction of two or more distributions

We shall assume that each sample size n_i is at least 10, unless all sample sizes $n_1, \ldots, n_r$ are approximately equal (in which event slighty smaller sample sizes can be tolerated). Calculate the median direction $\tilde{\theta}$ for the entire collection of N data points. For each $i = 1, \ldots, r$, let m_i be the number of values $\theta_{i1} - \tilde{\theta}, \ldots, \theta_{in_i} - \tilde{\theta}$ which are negative (where each $\theta_{ij} - \tilde{\theta}$ lies in the range $(-\pi, \pi)$), and let $M = m_1 + \cdots + m_r$. The test statistic is

$$P_r = \{N^2/[M(N-M)]\} \sum_{i=1}^{r} m_i^2/n_i - NM/(N-M) \tag{5.9}$$

and the hypothesis that the median directions are the same is rejected if P_r is too large.

Critical values. Compare P_r with the upper $100(1-\alpha)\%$ point of the χ^2_{r-1} distribution in Appendix A2.

The issue of comparing samples of vectors with samples of axes arises in the next example. In a testing situation, a simple rule is to double *all* the data (i.e. vectors *and* axes, modulo 360°) and carry out the test appropriate to vectorial data. For estimation, other changes are needed, as described later in this Section.

Example 5.7 For the data of Example 5.4, suppose we wish to compare the median axis of the groove marks with the median direction of the flutes. We get the median direction of the 36 *pooled doubled measurements* to be 34°, and m_1 (for grooves) $= 14, m_2$ (for flutes) $= 4$, whence $P_1 = 4.21$, slightly in excess of the upper 5% critical value of χ^2_1 (namely 3.84). So there is some evidence that the median axes of the groove and flute marks are different.

References and footnotes. The method of comparing vectorial and axial samples was suggested by Fisher & Powell (1989).

5.3.3 Estimation of the common median direction of two or more distributions

It is a straightforward matter to estimate the common median direction $\tilde{\mu}$ of two or more distributions, the estimate being just the median direction $\tilde{\theta}$ of the pooled sample values as used in §**5.3.2**. It is *not* so simple to calculate an interval estimate of $\tilde{\mu}$ (i.e. an assessment of the range of plausible values for $\tilde{\theta}$). We confine discussion to one special case, namely, the case in which the underlying distributions can be assumed to be identical. Here, we can regard all N data points in the pooled sample as being drawn from the same distribution, and the methods of §**4.4.2** can be applied. In other cases, methods based on the sample mean directions will usually be adequate: see §**5.3.5**.

5.3.4 Test for a common mean direction of two or more distributions

Our choice of method for comparing two or more sample directions is dictated by the sample sizes. With small sample sizes, we need to use bootstrap methods. With sufficiently large sample sizes, we can perform the test easily, using just the summary values $n, \bar{\theta}, \bar{R}$ and $\hat{\sigma}$ (cf. §**4.4.4**).

In some circumstances, it may be reasonable to suppose that the underlying distributions are *identical*, apart from possible differences in mean directions. For such data, an alternative method of analysis is given in §**5.5**. However, for large sample sizes ($n_i \geq 25$), the method described in §**5.3.4**(*b*) is simpler and just as good.

Both in small and in large samples, a decision has to be made as to whether the samples have comparable dispersions, leading to a choice between two procedures P and M below. If the largest of the sample dispersions $\hat{\delta}_1, \ldots, \hat{\delta}_r$ (cf. (2.28)) is no more than four times the smallest, the simpler method P can be used in calculating the test statistic; otherwise use method M. In either case, allowance is also made for differing sample sizes.

(a) Some $n_i \leq 25$. Use the general bootstrap method described in §**8.4.4**. If the underlying distributions can be taken to be symmetric about their mean directions, use the modified re-sampling scheme given in §**8.3.2** *Technique 2*.

> **Example 5.8** For the flutes and grooves data described in Example 5.4 and analysed in Example 5.7, consider comparing the mean axis of the grooves with the mean direction of the flutes. The first issue to decide is whether the samples have comparable dispersions. Since one sample is axial, we double both samples (modulo 360°), and calculate the sample dispersions (cf. (2.28)) for these derived samples, getting $\hat{\delta}_1 = 1.088$ and $\hat{\delta}_2 = 1.462$. As these are comparable, the analysis can be done using the P weights (5.15) in the formulae for the test statistic (5.10)–(5.13)). We obtain the test value $Y_2 = 0.0889$, from (5.13). The significance probability of this, using a bootstrap method with $B = 200$ resamples to estimate the bootstrap distribution, is 0.05. This is in agreement with the result in Example 5.7, namely that there is some evidence for differences in orientation of the two

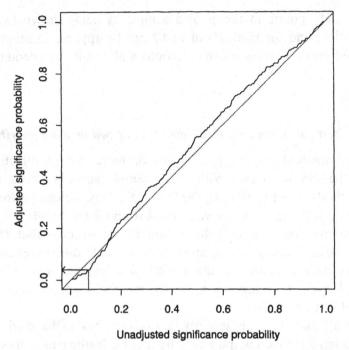

Fig. 5.8 Calibration plot to adjust significance probability for bootstrap comparison of mean trend of groove and tool marks and mean direction of flute marks of data in Figure 5.4. See Example 5.8.

data sets. Is it necessary to calibrate this test (*cf.* §**8.4.3**)? With these sample sizes, the answer is probably 'No': however, for illustration, we do so. The calibration plot is shown in Figure 5.8. It is essentially linear, and results in adjusting the significance probability from 0.07 to 0.04.

(b) Each $n_i \geq 25$. The method of testing depends on whether or not the samples have comparable dispersions.

Calculate the dispersion values $\widehat{\delta}_1, \ldots, \widehat{\delta}_r$ (*cf.* (2.28)), and let $\widehat{\delta}_{min}, \widehat{\delta}_{max}$ denote, respectively, the minimum and maximum values of $\widehat{\delta}_1, \ldots, \widehat{\delta}_r$. If $\widehat{\delta}_{max}/\widehat{\delta}_{min} \leq 4$ use method P below, else use method M.

Method P. If the dispersion weights $\widehat{\delta}_1, \ldots, \widehat{\delta}_r$ are comparable, in the sense that $\widehat{\delta}_{max}/\widehat{\delta}_{min} \leq 4$, calculate

$$\widehat{C}_P = \sum_{i=1}^{r} n_i \cos \widehat{\mu}_i, \qquad \widehat{S}_P = \sum_{i=1}^{r} n_i \sin \widehat{\mu}_i, \tag{5.10}$$

$$R_P = (\widehat{C}_P^2 + \widehat{S}_P^2)^{\frac{1}{2}} \tag{5.11}$$

and a weighted average $\widehat{\delta}_0$ as

$$\widehat{\delta}_0 = \sum_{i=1}^{r} n_i \widehat{\delta}_i / N \tag{5.12}$$

Then the test statistic is

$$Y_r = 2(N - R_P)/\widehat{\delta}_0 \tag{5.13}$$

The hypothesis of a common mean direction underlying the r samples is rejected if Y_r is too large.

Method M. If the dispersion weights $\widehat{\delta}_1, \ldots, \widehat{\delta}_r$ are *not* comparable, that is, $\widehat{\delta}_{max}/\widehat{\delta}_{min} > 4$, calculate

$$\widehat{C}_M = \sum_{i=1}^{r} (\cos \widehat{\mu}_i)/\widehat{\sigma}_i^2, \qquad \widehat{S}_M = \sum_{i=1}^{r} (\sin \widehat{\mu}_i)/\widehat{\sigma}_i^2 \tag{5.14}$$

$$R_M = (\widehat{C}_M^2 + \widehat{S}_M^2)^{\frac{1}{2}} \tag{5.15}$$

Then the test statistic is

$$Y_r = 2(\sum_{i=1}^{r} 1/\widehat{\sigma}_i^2 - R_M) \tag{5.16}$$

The hypothesis of a common mean direction underlying the r samples is again rejected if Y_r is too large.

Critical values. Compare Y_r with the upper $100(1-\alpha)\%$ point of the χ_{r-1}^2 distribution in Appendix A2.

> **Example 5.9** To compare the mean orientations of the termite mounds discussed in Examples 5.2 and 5.5, we carry out a formal test using the weights (5.16), because there is some evidence from Example 5.5 that the samples have differing dispersions; in fact, the sample dispersions range from about 0.10 to 0.63 once the data are converted to vectors. We get the values $\sum 1/\widehat{\sigma}_i^2 = 4275.16$ and $R_M = 4229.34$, whence $Y_{14} = 91.6$. This is an extremely unlikely value of the χ_{13}^2 distribution. So there is strong evidence of differences between the mean orientations of the termitaria.

References and footnotes. The large-sample test based on Y_r is due to Watson (1983, pp. 146–7). The bootstrap test was proposed by Fisher & Hall (1991a); some formulae are in error in that paper, although not so as to alter the results of bootstrap testing, only of large-sample testing.

5.3.5 Estimation of the common mean direction of two or more distributions

Suppose that we wish to form a pooled estimate of the common mean direction μ, based on the separate-sample estimates $\hat{\mu}_1, \ldots, \hat{\mu}_r$. As with testing for differences between mean directions in §**5.3.4**, the method of forming this pooled estimate should take account of

- possibly differing sample sizes; and
- possibly differing dispersions, or spreads in the samples.

As well, we need to use different methods for small sample sizes ($n_i < 25$) than for large samples. One simple case can be dismissed immediately: if, on the basis of, say, the test in §**5.5**, we decide that the samples all come from the *same* distribution, then the r samples can all be combined into a super-sample of size $N = n_1 + \cdots + n_r$, and the methods in §**4.4.4** applied.

We describe, below, two different methods of weighting the individual mean directions $\hat{\mu}_i$, depending on variation amongst the sample circular dispersions $\hat{\delta}_1, \ldots, \hat{\delta}_r$. As a rough guide, the simpler weighting method P can be used provided that the largest $\hat{\delta}_i$ is not more than four times the size of the smallest; otherwise use method M. (These estimation methods correspond to methods P and M in §**5.3.4**.)

(a) **Some** $n_i < 25$. Use the bootstrap method described in §**8.3.4**. If the samples can be assumed to be from distributions with comparable spreads $\delta_1, \ldots, \delta_r$ (largest not greater than four times the smallest), use the P weights defined by (5.22), otherwise the M weights defined by (5.26).

> **Example 5.10** Figure 5.9 shows the walking directions of three groups of monocular ants which can see only the pattern of polarised light in the sky. (The data are listed in Appendix B10.) The control group (a) was studied in Example 4.15. Ants in the two treatment groups were being tested for interocular transfer, group $b1$ (32 ants) having their naive eyes occluded while being trained to the $0°$ compass direction, and group $b2$ (18 ants) having their naive eyes occluded for the duration of the experiment. Comparison of the data displays does not suggest any difference in the mean directions of the three groups, although the dispersions differ somewhat. So to calculate a pooled estimate of the common mean direction, it is appropriate to use method M for weighting the different samples. We get the pooled estimate of the mean direction as $3.8°$, with 95% confidence interval $(352.9°, 13.9°)$.

(b) **All** $n_i \geq 25$. We begin by describing some general weighting schemes, which we then specialize to the two methods P and M mentioned above.

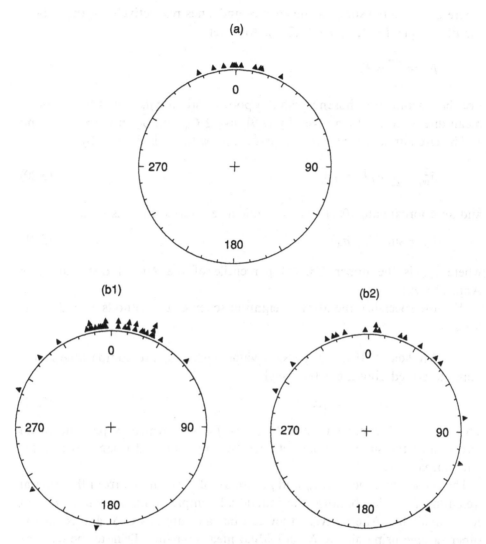

Fig. 5.9 Directions of three groups of ants, from an experiment to test for interocular transfer. (a) Control group. (b1) Naive eyes occluded while being trained. (b2) Naive eyes occluded for the lifetime of the experiment. See Example 5.10.

Let $w_1 \ldots, w_r$ be any positive weights which sum to 1, and let

$$\bar{C}_w = \sum_{i=1}^{r} w_i \bar{R}_i \cos \hat{\mu}_i \equiv \sum_{i=1}^{r} w_i C_i / n_i \qquad (5.17)$$

$$\bar{S}_w = \sum_{i=1}^{r} w_i \bar{R}_i \sin \hat{\mu}_i \equiv \sum_{i=1}^{r} w_i S_i / n_i \qquad (5.18)$$

where C_i, S_i are the sums of the cosines and sines respectively, of the data in the ith sample, $i = 1, \ldots, r$ (cf. (2.7)). Also, let

$$\bar{\rho}_w = \sum_{i=1}^{r} w_i \bar{R}_i \tag{5.19}$$

For these arbitrarily chosen weights, a pooled estimate $\hat{\mu}_w$ say of the common mean direction μ is then given by (2.9), using $\hat{C}_w$ and $\hat{S}_w$ in place of C and S. The circular standard error $\hat{\sigma}_w$ associated with $\hat{\mu}_w$ is defined by

$$\hat{\sigma}_w^2 = \sum_{i=1}^{r} w_i^2 \bar{R}_i^2 \hat{\sigma}_i^2 / \bar{\rho}_w^2 \tag{5.20}$$

and an approximate $100(1-\alpha)\%$ confidence interval for μ is then

$$\hat{\mu}_w \pm \sin^{-1}(z_{\frac{1}{2}\alpha}\hat{\sigma}_w) \tag{5.21}$$

where $z_{\frac{1}{2}\alpha}$ is the upper $100(\frac{1}{2}\alpha)$ percentile of the $N(0,1)$ distribution in Appendix A1.

We now specialise the above weighting schemes to methods P and M, as follows.

Method P. If the dispersion values $\hat{\delta}_1, \ldots, \hat{\delta}_r$ are comparable in the sense described above, use the weights

$$w_i = n_i/N, \quad i = 1, \ldots, r \tag{5.22}$$

This leads to the pooled estimate $\hat{\mu}_w = \hat{\mu}_P$ say, which is just the mean direction of the super-sample obtained by combining all r samples to form one sample of size N.

The pooled qualities $\bar{C}_w, \bar{S}_w$ and $\bar{\rho}_w$ are usually calculated from the general weighting formulae because the individual sample summary values $\hat{\mu}_i$ and $\bar{R}_i$ are available. Alternatively, they can be calculated as C, S, and $\bar{R}$ for the super-sample using all the N individual measurements. Denote the circular standard error by $\hat{\sigma}_P$. An approximate $100(1-\alpha)\%$ confidence interval for the unknown common mean direction μ is then

$$\hat{\mu}_P \pm \sin^{-1}(z_{\frac{1}{2}\alpha}\hat{\sigma}_P) \tag{5.23}$$

Method M. If the $\hat{\delta}_i$ values differ substantially, define the quantities

$$v_i = (\bar{R}_i \hat{\sigma}_i^2)^{-1}, \quad i = 1, \ldots, r \tag{5.24}$$

and

$$v = \sum_{i=1}^{r} v_i \tag{5.25}$$

Table 5.2. *Summary statistics for the three Bulgoo samples of cross-bed azimuths, required to compare the three sample mean directions and, if appropriate, to calculate a pooled estimate of the common mean direction.*

	Sample			Pooled estimation method	
Statistic	1	2	3	P	M
$\hat{\mu}_i$	228.1°	247.6°	235.5°	238.3°	242.1°
n_i	40	30	30	100	100
$\bar{R}_i$	0.4049	0.7828	0.6088	0.5736	0.6537
$\hat{\sigma}_i$	0.2968	0.1314	0.2012	0.1172	0.1032

Then the weights are

$$w_i = v_i/v, \quad i = 1,\ldots,r \tag{5.26}$$

This leads to an estimate, $\hat{\mu}_M$ say, of μ, with associated circular standard error $\hat{\sigma}_M$ and $100(1-\alpha)\%$ confidence interval

$$\hat{\mu}_M \pm \sin^{-1}(z_{\frac{1}{2}\alpha}\hat{\sigma}_M) \tag{5.27}$$

(This is fractionally different from the interval $\hat{\mu}_w \pm \sin^{-1}(z_{\frac{1}{2}\alpha}\hat{\sigma}\bar{\rho}_w/\bar{R}_w)$ defined by Fisher and Lewis (1983), where $\bar{R}_w^2 = \bar{C}_w^2 + \bar{S}_w^2$. However, the difference is typically negligible in practice, and the form recommended here is simpler and directly analogous to that used for a single sample.)

> **Example 5.11** Consider the three samples of cross-bed azimuths of Example 5.1. The raw data plots in Figure 5.1 suggest little evidence that the mean directions differ, and this is confirmed by using the test (5.14) and the M-weights (5.16). We shall calculate a pooled estimate of the common mean direction using the two methods. Method P yields a pooled estimate of 238.3° with a circular standard error of 0.1172 and so a 95% confidence interval of (225.0°, 251.6°). Method M yields a pooled estimate of 242.1° with a circular standard error of 0.1032 (so that it provides a marginally more precise estimate in this case) and 95% confidence interval (230.3°, 2253.9°) for the overall mean direction. The summary statistics required to perform this pooling, together with the resulting summary values, are shown in Table 5.2.

References and footnotes. The large-sample methods for combining sample mean directions are due to Fisher & Lewis (1983) and Watson (1983). The bootstrap methods for calculating confidence regions for the mean direction were suggested by Fisher & Powell (1989).

5.3.6 *Test whether two or more distributions are identical*

In §**5.3.2** and §**5.3.4**, we considered tests of the hypothesis that the distributions underlying the observed samples of data had the same reference direction. Occasionally it is of interest to test rather more than this, namely that the distributions are identical (a so-called test of *homogeneity*). If we have fitted von Mises models to the samples, then it is simply a matter of testing whether the mean directions and dispersions are the same, as is done in the next section. However, if the von Mises model is not satisfactory – for example, if the data sets exhibit multimodality – we need a more general method.

Unimodal samples can be compared informally by using the Q–Q plots described in §**5.2** to investigate the shape differences relative to sample mean directions. However, for multimodal distributions, these plots can be difficult to interpret. Such difficulties do not generally apply when a formal test is used.

To carry out a formal test we shall use the notion of circular ranks cf. §**2.3.3**). Considering all $N = n_1 + \cdots + n_r$ data values as a single sample, calculate their circular ranks, and let γ_{ij} denote the circular rank of θ_{ij} among all the data. For each sample $i, i = 1, \ldots, r$, calculate

$$C_i = \sum_{j=1}^{n_i} \cos \gamma_{ij}, \quad S_i = \sum_{j=1}^{n_i} \sin \gamma_{ij} \tag{5.28}$$

and hence the test statistic

$$W_r = 2 \sum_{i=1}^{r} (C_i^2 + S_i^2)/n_i \tag{5.29}$$

The null hypothesis that the distributions are identical is rejected if W_r is too large.

Critical values. If any of the sample sizes n_i is less than 10, a randomisation test procedure can be used as described in §**8.5**, based on the criterion W_r. Suppose we have performed 1000 randomisations of the N data points, and calculated the statistics $w_1, \ldots, w_{1000}$ corresponding to each of these. Let w_0 be the value of W_r for our actual sample. To test at the $100(1 - \alpha)\%$ level, calculate

$$m = \text{largest integer not exceeding} \quad 1000(1 - \alpha) + 1.$$

The hypothesis is rejected if $w_0 > w_m$. Alternatively, if w_0 exceeds M of the 1000 randomisation w-values, the significance probability of w_0 is $(1000 - M)/1000$.

If $n_i \geq 10$, $i = 1, \ldots, r$, compare W_r with the upper $100(1-\alpha)\%$ point of the χ^2_{2r-2} distribution in Appendix A2.

> **Example 5.12** Consider the wind direction data described in Example 5.3. From the data plots in Figure 5.3, the data collected during Summer appear rather different from those in the other seasons, so we restrict attention to comparing the Winter, Spring and Autumn. Where two or more data points have the same value, each datum has been assigned the average of the two or more associated ranks. The value of W_3 is 4.16, corresponding to a significance probability of about 0.38 for a χ^2_4 variate, so there is little evidence that the wind distributions in these three seasons differ.

References and footnotes. There are numerous nonparametric tests for testing the hypothesis that two or more samples have been drawn from identical distributions. This is due, in part, to the close relationship between r-sample tests of homogeneity and one-sample tests of uniformity (see e.g. Beran 1969b). The test recommended here is an r-sample extension by Mardia (1972b) of the two-sample Uniform Scores discussed by Wheeler & Watson (1964), following a suggestion by J.L. Hodges Jr. For a general discussion of such tests and further references see Mardia (1972b), Rao (1984) and Upton & Fingleton (1989). Table 9.32 of the last reference contains a helpful guide to choice of the best test in a particular situation. Here, a decision has been made to recommend a test which is easy to implement and which performs reasonably in a variety of situations. Mardia & Spurr (1973) studied r-sample tests for data drawn from l-modal distributions, of which the simplest example would be the case of axial data ($l = 2$) for which the doubled data are vectors. The recommendation in this book is to convert the data to vectors – $\theta_{ij} \rightarrow l\theta_{ij}$ – and analyse the resulting vectors, as it appears to work satisfactorily and requires no new methodology beyond what is already available.

5.4 Analysis of two or more samples from von Mises distributions

5.4.1 Introduction

With the additional assumption that each sample of data is drawn from a von Mises distribution (cf. §3.3.6 and the goodness-of-fit procedures in §4.5.3) we can compare dispersions of the samples as well as their mean directions. As in §5.3, the choice of methods used to compare mean directions will depend on whether or not the dispersions *are* comparable, and this can be decided using the formal test described in §5.4.3 below. We begin with analysis of mean directions.

5.4.2 Test for a common mean direction of two or more von Mises distributions

The methods in §**5.4.4** can be used to decide whether or not the assumption $\kappa_1 = \cdots = \kappa_r$ is reasonable. How we proceed depends on this decision, and on the sample sizes.

Generally, the approach here is similar to what was used in the more general setting of §**5.3.4** (in which mean directions were compared, without the von Mises assumption). From §**4.4.5**, the standard error of the mean direction of the ith von Mises sample takes the particular form

$$\hat{\sigma}_i = 1/(n_i \bar{R}_i \hat{\kappa}_i)^{\frac{1}{2}} \tag{5.30}$$

(cf. (4.42)), the associated sample dispersion is

$$\hat{\delta}_i = 1/(\bar{R}_i \hat{\kappa}_i)^{\frac{1}{2}} \tag{5.31}$$

and the test statistic for Method M defined in (5.16) specialises to

$$Y_r = 2(\sum_{i=1}^{r} n_i \bar{R}_i \hat{\kappa}_i - R_M) \tag{5.32}$$

(a) Concentration parameters $\kappa_1, \ldots, \kappa_r$ not all equal

Unless the sample sizes are all large, and each estimate $\hat{\kappa}_i$ (cf. (4.41)) exceeds 2, the best approach is a bootstrap method.

(a1) Unequal concentration parameters, at least one $n_i < 25$ or $\hat{\kappa}_i < 2.0$

Use the general bootstrap method described in §**8.4.4**, with re-sampling from the von Mises distributions $VM(0, \hat{\kappa}_i)$, $i = 1, \ldots, r$, and based on the statistic Y_r in (5.32).

> **Example 5.13** Consider the two samples of cross-bed measurements dis-cussed in Example 5.6. The analysis suggested that the samples were drawn from distributions with different shapes – in this context, different κ – but not necessarily with differing mean directions, which issue we now pursue. Analysis of each data set using the methods of §**4.5.3** indicates that a von Mises model fits each set adequately: the von Mises Q–Q plots are shown in Figure 5.10. Summary values for each data set are as follows:
>
> $n_1 = 25,\quad \hat{\mu}_1 = 150.3°,\quad \hat{\kappa}_1 = 0.53$
> $n_2 = 104,\quad \hat{\mu}_2 = 122.3°,\quad \hat{\kappa}_2 = 0.89$
>
> In the light of the graphical evidence for differing dispersions, and the low values of the estimates $\hat{\kappa}_1$ and $\hat{\kappa}_2$, we use a bootstrap method with the statistic given by (5.32). We get the value of Y_2 to be 0.88, corresponding to

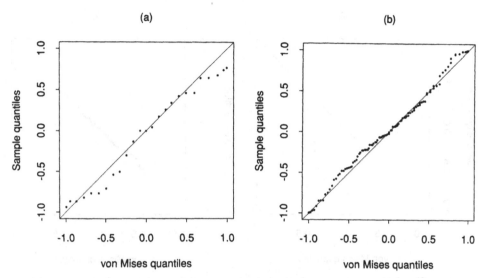

Fig. 5.10 Figure 5.10 von Mises Q–Q plots to check the goodness-of-fit of von Mises models for the data in Figure 5.6. In both cases, the fit seems adequate. See Example 5.13.

a significance probability of about 0.31, so there is little evidence that the mean directions differ.

(a2) Unequal concentration parameters, each $n_i \geq 25$, each $\hat{\kappa}_i \geq 2.0$

Use the statistic Y_r in (5.32).

Note: The tests in *(a1)* and *(a2)* also apply even if $\kappa_1, \ldots, \kappa_r$ are *known* (and unequal).

Example 5.14 Consider data sets 10 and 11 of the termitaria described in Examples 5.2 and 5.5. Since the data are axial, to evaluate the von Mises model we convert them to vectors first. von Mises Q–Q plots for the vector data are shown in Figure 5.11, and indicate that fitting such models is reasonable in each case. Summary values for each data set are as follows (noting that the estimates $\hat{\kappa}_1$ and $\hat{\kappa}_2$ refer to the vector data):

$$n_1 = 50, \quad \hat{\mu}_1 = 172.5°, \quad \hat{\kappa}_1 = 9.8$$
$$n_2 = 100, \quad \hat{\mu}_2 = 0.6°(\equiv 180.6°), \quad \hat{\kappa}_2 = 2.1$$

We get $Y_2 = 9.2$, corresponding to a significance probability less than 0.01, so there is strong evidence for a difference between the two mean orientations.

(a)   (b)

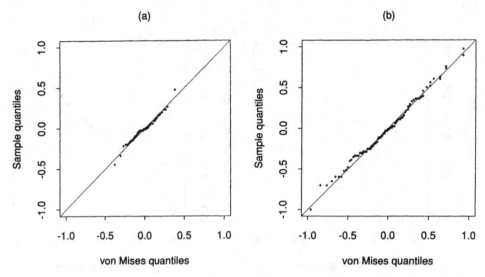

Fig. 5.11 von Mises Q–Q plot to check the goodness-of-fit of von Mises models for data sets 10 and 11 in Figure 5.2. In both cases, the fit seems adequate. See Example 5.14.

(b) Concentration parameters equal ($\kappa_1 = \cdots = \kappa_r = \kappa$)

Let

$$\widetilde{\kappa} = \text{median value of } \widehat{\kappa}_1, \ldots, \widehat{\kappa}_r \tag{5.33}$$

(b1) Equal concentration parameters, $\widetilde{\kappa} \geq 2.0$

Let R be the resultant length of all N data points, computed either directly, or from the sample summary values using

$$R = \left[\left(\sum_{i=1}^{r} R_i \cos \widehat{\mu}_i \right)^2 + \left(\sum_{i=1}^{r} R_i \sin \widehat{\mu}_i \right)^2 \right]^{\frac{1}{2}} \tag{5.34}$$

The test statistic is

$$T_r = (N - r) \left(\sum_{i=1}^{r} R_i - R \right) \bigg/ \left[(r - 1) \left(N - \sum_{i=1}^{r} R_i \right) \right] \tag{5.35}$$

and the hypothesis that $\mu_1 = \cdots = \mu_r$ is rejected if T is too large.

Critical values. Compare T with the upper $100(1 - \alpha)\%$ point of the $F_{r-1, N-r}$ distribution. This distribution is available in all books of standard statistical tables, e.g. Pearson & Hartley (1970, Table 18).

Example 5.15 Consider data sets 5 and 14 of the termitaria described in Examples 5.2 and 5.5. As in Example 5.14, to evaluate the von Mises model

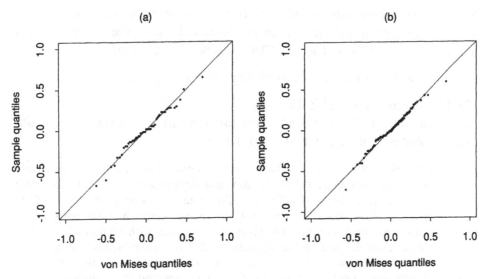

(a)   (b)

Sample quantiles

von Mises quantiles

Fig. 5.12 von Mises Q–Q plot to check the goodness-of-fit of a von Mises models for data sets 5 and 14 in Figure 5.2. In both cases, the fit seems adequate. See Example 5.15.

we convert them to vectors first. von Mises Q–Q plots for the vector data are shown in Figure 5.12, and indicate that fitting such models is reasonable in each case. Summary values for each data set are as follows (the estimates $\hat{\kappa}_1$ and $\hat{\kappa}_2$ referring to the vector data):

$$n_1 = 50, \quad \hat{\mu}_1 = 173.0°, \quad \hat{\kappa}_1 = 3.9$$
$$n_2 = 92, \quad \hat{\mu}_2 = 179.5°(\equiv 180.5°), \quad \hat{\kappa}_2 = 5.4$$

In Example 5.19 below, it is shown that the assumption that $\kappa_1 = \kappa_2$ is entirely reasonable. Making this assumption, and using (5.35), $T_r = 6.31$ with a significance probability of about 0.01. We conclude that there is strong evidence for a difference between the two mean orientations.

(b2) Equal concentration parameters, three or more samples,
   $1 < \tilde{\kappa} < 2.0$

Calculate T_r as in (5.35), and

$$T'_r = [1 + 3/(8\tilde{\kappa})]\,T_r, \tag{5.36}$$

and proceed as in *(b1)* using T'_r instead of T_r as the test statistic.

(b3) Equal concentration parameters, all other cases

Use a randomisation test, as described in §**8.5**, based on the criterion T in (5.35). Suppose we have performed 1000 random allocations of the

N data points into samples of sizes $n_1, \ldots, n_r$, and calculated the statistics $T_1, \ldots, T_{1000}$ corresponding to each of these. Let T_0 be the value of T for our actual sample. To test at the $100(1 - \alpha)\%$ level, calculate

$$m = \text{largest integer not exceeding } 1000(1 - \alpha) + 1$$

The hypothesis is rejected if $T_0 > T_m$.

Alternatively, if T_0 exceeds M of the randomisation T–values, the significance probability of T_0 is $(1000 - M)/1000$.

> **Example 5.16** Figure 5.13 shows small samples of orientations of core samples (axial data) collected at five sampling locations in the Pacheco Pass area of the Diablo Range, California. (The data are listed in Appendix B17.) In view of the plots and the small sample sizes, we shall assume that the data come from bipolar von Mises distributions with comparable dispersions. Consider testing the hypothesis that the samples have been drawn from distributions with the same mean direction – that is, from the same distribution. Because of the small sample sizes and large dispersions, we use a randomisation test. Summary statistics for the data of each sample, transformed to vectors, are:
>
> | n_i : | 8 | 14 | 11 | 9 | 6 |
> |---------|------|------|------|------|------|
> | R_i : | 5.91 | 5.23 | 4.40 | 6.63 | 4.69 |
>
> Also, $N = 48, R = 14.94$, leading to $T_0 = 6.06$. Using a randomisation test as described above ($M = 1000$) results in a significance probability of 0.001, so there is strong evidence that the mean orientations of the cores differ.

References and footnotes. The large-sample methods for the case of unequal dispersions are due to Watson (1983, Chapter 4), and the corresponding small-sample bootstrap methods to Fisher & Hall (1990a) (note corrections in Footnote to §5.3.4). In the case of equal dispersions, the statistic T_r is due to Watson & Williams (1956), and the modified form T_r' was given by Mardia (1972a) based on an approximation in Stephens (1969a) and other recommendations by Stephens. Various tests have been suggested for the equal-dispersion case, because all parametric test statistics have distributions depending on the value of κ, when κ is small (e.g. $\kappa < 2$): see Upton & Fingleton (1989 §9.8) for a review and guide to the relevant literature. The approach taken here has been to make a recommendation on the grounds of simplicity: namely, that in situations when the sample sizes or estimated value of κ are such as to make the Watson & Williams test (or its simple modification) inappropriate, then a randomisation test should be used. Mardia & Spurr (1973) studied r-sample tests for data drawn from l-modal von Mises distributions, of which the simplest example would be axial data for which the doubled data are von Mises. As with the comment

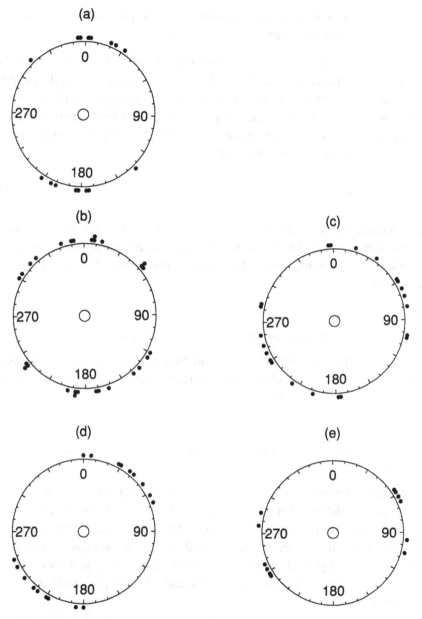

Fig. 5.13 Samples of orientations of rock cores taken from five sampling sites within the Pacheco Pass area, Diablo Range, California. See Example 5.16.

at the end of §**5.3.6**, the recommendation here is to convert the data to vectors – $\theta_{ij} \to l\theta_{ij}$ – and analyse the resulting vectors. Mardia & El-Atoum (1976) investigated a Bayesian approach for known κ.

5.4.3 Estimation of the common mean direction of two or more von Mises distributions

As with the preceding section on comparing mean directions of von Mises samples, the approach we adopt here is similar to what was used in the more general setting of §5.3.5 (in which mean directions were combined, without the von Mises assumption). The sample circular standard errors and circular dispersions are now given by (5.30) and (5.31) respectively. From the values $v_1, \ldots, v_r$ in (5.32) we define the weights

$$w_i = v_i/(v_1 + \cdots + v_r) \tag{5.37}$$

which replace those given in (5.26) for Method M. How we combine the separate sample mean directions will depend on the sample sizes and circular dispersions – or equivalently, the sample estimates of the concentration parameters.

(a) Concentration parameters $\kappa_1, \ldots, \kappa_r$ not all equal

Unless the sample sizes are all large, and each estimate $\hat{\kappa}_i$ (cf. (4.41)) exceeds 2, the best approach to calculating a confidence region for the common mean direction is a bootstrap method.

(a1) Unequal concentration parameters, at least one $n_i < 25$ or $\hat{\kappa}_i < 2.0$

Use the general bootstrap method described in §8.3.4, with re-sampling from the von Mises distributions $VM(0, \hat{\kappa}_i)$, $i = 1, \ldots, r$, to estimate the common mean direction and an associated confidence interval. Separate sample bootstrap mean directions are combined using the approach in §5.3.5(b) and the circular standard errors and circular dispersions given in (5.30) and (5.31) respectively. If Method M is selected, the weights defined in (5.37) should be used.

> **Example 5.17** The two samples of cross-bed measurements analysed in Examples 5.6 and 5.13 were found to have comparable mean directions. We shall estimate the common mean direction μ and calculate a confidence interval for it. Using the weights defined in (5.37) yields the estimate $\hat{\mu} = 124.9°$. Since $\hat{\kappa}_1$ and $\hat{\kappa}_2$ are both small, we use a bootstrap method to compute the confidence region. A 95% confidence interval based on 200 bootstrap resamples is 122.6°, 128.1°).

(a2) Unequal concentration parameters, each $n_i \geq 25$, each $\widehat{\kappa}_i \geq 2.0$

Use the method in §**5.3.5**(b), except with circular standard errors and circular dispersions given by (5.30) and (5.31) respectively. If Method M is employed, the weights $w_1, \ldots, w_r$ given by (5.36) should be used.

> **Example 5.18** Consider data sets 10 and 15 of the termitaria described in Examples 5.2 and 5.5. Since the data are axial, to evaluate the von Mises model we convert them to vectors first. von Mises Q–Q plots for the vector data are shown in Figures 5.11 and 5.12, and indicate that fitting such models is reasonable in each case. Summary values for each data set are as follows (noting that the estimates $\widehat{\kappa}_1$ and $\widehat{\kappa}_2$ refer to the vector data):
>
> $$n_1 = 50, \quad \widehat{\mu}_1 = 172.5°, \quad \widehat{\kappa}_1 = 9.8$$
> $$n_2 = 50, \quad \widehat{\mu}_2 = 173.0°, \quad \widehat{\kappa}_2 = 3.9$$
>
> The pooled estimate of the mean orientation is 172.6°, with estimated standard error (for the corresponding vector direction) of 0.0396, giving a 95% confidence interval of $(170.4°, 174.9°)$.

(b) Concentration parameters equal $(\kappa_1 = \cdots = \kappa_r = \kappa)$

If it is reasonable to assume that equality of the concentration parameters $\kappa_1, \ldots, \kappa_r$ (cf. §**5.4.4**) then all data may be pooled into a super-sample of N data points, and the methods in §**4.5.5** applied.

> **References and footnotes.** The large-sample methods for the case of unequal dispersions are due to Watson (1983, Chapter 4). Mardia & El-Atoum (1976) investigated a Bayesian approach with a common known value of κ.

5.4.4 Test for equality of concentration parameters of two or more von Mises distributions

Consider testing the null hypothesis $\kappa_1 = \cdots = \kappa_r = \kappa$ *say* against the alternative that at least one κ_i differs from the others. For sample number i, calculate the sample mean direction $\widehat{\mu}_i$ and then

$$d_{ij} = |\sin(\theta_{ij} - \widehat{\mu}_i)|, \quad j = 1, \ldots, n_i, \quad \bar{d}_i = \sum_{j=1}^{n_i} d_{ij}/n_i \tag{5.38}$$

and

$$\bar{d} = \sum_{i=1}^{r} n_i \bar{d}_i / N \tag{5.39}$$

The test statistic is

$$f_r = \frac{(N-r)\sum\limits_{i=1}^{r} n_i(\bar{d}_i - \bar{d})^2}{(r-1)\sum\limits_{i=1}^{r}\sum\limits_{j=1}^{n_i}(d_{ij} - \bar{d}_i)^2} \tag{5.40}$$

and the hypothesis that $\kappa_1 = \cdots = \kappa_r$ is rejected if f_r is too large.

Critical values. The method of assessing the value of f_r depends on sample sizes and dispersions. First, obtain $\tilde{\kappa}$ from (5.33).

All samples sizes ≥ 10, $\tilde{\kappa} \geq 1$. Compare f_r with the upper $100(1 - \alpha)\%$ point of the $F_{r-1,N-r}$ distribution (see e.g. Pearson & Hartley 1970, Table 18).

All other cases. Centre each sample i at its sample mean direction $\hat{\mu}_i$, and then use a randomisation test (cf. §**8.5**) based on allocating the N centred values to the r samples, and calculating f_r for each of the randomised allocations. This is an approximate method.

Note: When just two samples are being compared, two tests are possible, a one-sided test (e.g. $H_0 : \kappa_1 \leq \kappa_2$ against $H_1 : \kappa_1 > \kappa_2$) or a two-sided test ($H_0 : \kappa_1 = \kappa_2$ against $H_1 : \kappa_1 \neq \kappa_2$). In the latter case, the samples should be labelled Sample 1 and Sample 2 in such a way that $f_r \geq 1$; then the value of f_r should be compared with the upper $100(\frac{1}{2}\alpha)$ percentile of the F-distribution or randomisation distribution (as appropriate) rather than the upper 100α percentile. (Alternatively, if the significance probability of f_r has been found, it should be doubled.)

Example 5.19 For the two samples of orientations of termite mounds studied in Example 5.15, we shall test the hypothesis that $\kappa_1 = \kappa_2$, which was assumed in that analysis. From (5.40), $f_2 = 1.01$, which is not an extreme value of the $F_{1,140}$-distribution. We conclude that there is no evidence in the data to suggest that κ_1 and κ_2 differ.

References and footnotes. The test based on f_r is due to Fisher (1986). Other tests recommended in the literature have tended to be sensitive to the presence of outliers and to the von Mises assumption. A review of these tests and references to the literature are given by Upton & Fingleton (1989, §9.8).

5.4.5 Estimation of the common concentration parameter of two or more von Mises distributions

We assume that the mean directions $\mu_1, \ldots, \mu_r$ of the samples are unknown, or at least, that they cannot be presumed equal, for otherwise the N data points could be treated as a single sample (in the former case, after centering each sample at its known mean direction) and the methods of §4.5.5 applied.

If the common value κ, say, of $\kappa_1, \ldots, \kappa_r$ is small, κ will be estimated by bootstrap methods, otherwise more directly. The median value $\tilde{\kappa}$ in (5.33) can be used to make this decision.

(a) $\tilde{\kappa} < 2.0$. Estimate κ as in *(b)* below. Then compute a confidence interval for κ by using the bootstrap methods in §8.3.4 to get sets of bootstrap resultant lengths $R_1^*, \ldots, R_r^*$ for each of which a value R_{tot}^* can be calculated from (5.41) and hence a bootstrap estimate $\hat{\kappa}^*$.

(b) $\tilde{\kappa} \geq 2.0$. Let

$$R_{tot} = R_1 + \cdots + R_r \tag{5.41}$$

and estimate κ as in §4.5.5 using R_{tot} instead of R. A confidence interval for κ can be computed as in §4.5.5*(ii)(b)*, with $N - R_{tot}$ replacing $n - R$ throughout, and $N - r$ replacing $n - 1$ throughout.

5.5 Analysis of data from more complicated experimental designs

In contrast to the situation with linear data, very little in the way of statistical methodology is available for analysing data from experiments in which a circular response variable has been measured at various levels of two or more explanatory variables. The following discussion is confined to a brief survey of the literature.

The analysis of the preceding section was concerned with the so-called *one-way classification* design, in which a (von Mises) variate Θ is measured in each of r 'levels' of a single classification. In the language of the previous section, the observations of Θ in the ith level constituted the ith sample. More generally, we can have $s \geq 2$ classifications, each with several levels (a *cross-classified* or *factorial* experimental layout) with a sample of several observations made at each combination of levels. With this increased level of complexity, the issue of equal dispersions from sample to sample becomes even more critical, and careful diagnostic plots are required to check this aspect of the model.

Stephens (1982) presented an extension of the procedure in §5.4.2*(b)* for analysing cross-classified experiments with constant dispersion. A similar suggestion was made by Harrison *et al.* (1986), who noted the limitations of the test in terms of valid range of κ and with the interpretation of interactions. Harrison *et al.* (1986) and Harrison & Kanji (1988) then developed alternative methods for analysing randomised block designs and two-way classifications which are claimed to work for κ as small as 1. The methods are based on squares of component resultant lengths, and so may be sensitive to outliers in the data, and to departures from the von Mises assumption.

Graves (1979) studied randomisation methods for analysing discrete orientation data collected from an experiment based on a two-way classification. In this experiment, animals were placed at the centre of a regular n-sided polygon and a signal was emitted from one of the vertices and the direction in which the animal subsequently moved was recorded as that of the closest vertex. The randomisation approach was used more generally, to analyse data arising from factorial designs, in a series of papers by Underwood & Chapman (1985, 1989, 1992), Chapman (1986) and Chapman & Underwood (1992) in which they studied the movements of small animals in the presence of a variety of factors. Randomisation methods (cf. §**8.5**) were used to assess the significance of different effects, rather than relying on the von Mises assumption to obtain the null distributions of the statistics. The difficulty still remains of allowing for possible variation from group to group. In fact, as will be seen in Chapter 6 in the more general regression context, it is not uncommon that the dispersion be dependent on the value (here, level) of the explanatory variables. Much more experience with a range of experiments is needed to develop a coherent general approach in which the assumptions made in the statistical modelling can be properly evaluated.

6

Correlation and regression

6.1 Introduction

Whereas the previous chapters have been concerned with statistical inference for a single sample of data (typically, relating to the reference direction or dispersion) and with comparison of such quantities across several samples, this chapter and the next are concerned with relationships between random variables, or the dependence of a circular random variable on another variable. Examples of the sorts of problems to be studied in this chapter are:

Example 6.1 Figure 6.1 shows 19 measurements of wind direction and ozone concentration taken at 6.00am at four-day intervals between April 18th and June 29th, 1975, at a weather station in Milwaukee. (The data are listed in Appendix B18.) It is of interest to ascertain whether there is any association between ozone concentration and wind direction.

Example 6.2 Figure 6.2 shows the orientations (θ_i) of the nests of 50 noisy scrub birds along the bank of a creek bed, together with the corresponding directions (ϕ_i) of creek flow at the nearest point to the nest. (The data are listed in Appendix B19.) The separate data sets for the nest orientations and the creek flows are shown, together with a joint data plot. The directions of creek flow suggest that the creek has a small bend in the vicinity of data collection. The joint data plot consists of a scatterplot of the values $(\theta_i, \phi_i), i = 1, \ldots, n$, together with a duplicate set $(\theta_i + 360°, \phi_i + 360°), i = 1, \ldots, n$ (shown here in radians) to avoid difficulties due to transferring the data to the x-y plane. It is easier to view the pairs of circular data this way, rather than as points on a projection of the surface of a torus. If each ϕ_i differs from the corresponding θ_i by a fixed rotation ψ say, the plot will be a straight line with slope 45°.

The data were gathered as part of an experiment to investigate the relationship of the birds' nests to their physical surroundings. One question of interest is whether there is a dependence of nest orientation on creek bed orientation.

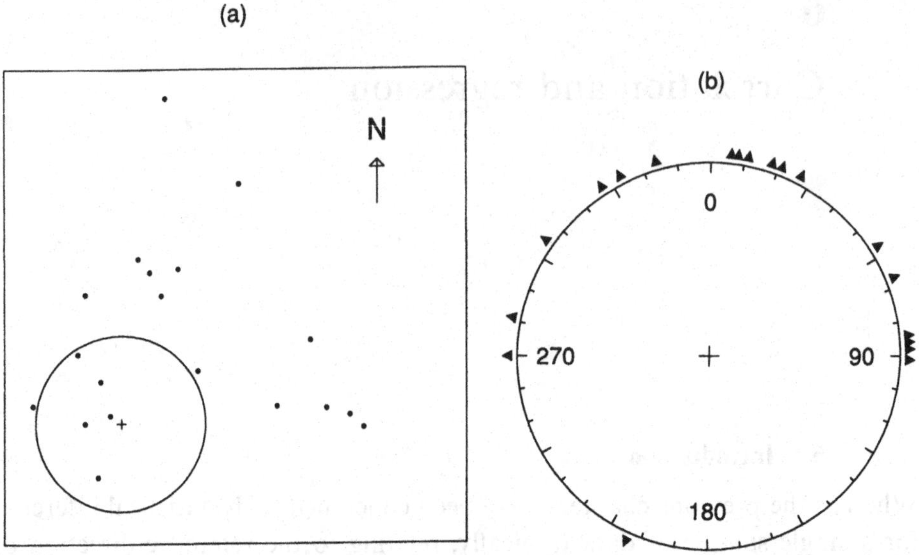

(a)

(b)

Fig. 6.1 Measurements of wind direction and ozone concentration at four-day intervals at a weather station in Milwaukee. (a) Joint data plot, with the ozone concentrations plotted as distances from the centre of the circle. (b) Marginal plot of wind directions. See Example 6.1.

Example 6.3 Figure 6.3 shows some data drawn from a series of experiments by Chapman (1986), Chapman & Underwood (1992), and Underwood & Chapman (1985, 1989, 1992) on distances and directions moved by small blue periwinkles, *Nodilittorina unifasciata*, after they had been transplanted downshore from the height at which they normally live. (The data are listed in Appendix B20.) The results of experiments at two different locations have been combined for the purposes of this example. The positions shown in the Figure correspond to a total of 31 animals, 15 of which were measured one day after transplantation and the other 16 four days after; plots of the separate sets indicate comparable behaviour. For each periwinkle, the direction shown is the compass direction relative to its point of release; the arrow shows the approximate direction (275°) of the sea.

Loosely speaking, regression analysis aims to model the mean value of a response variable in terms of one or more explanatory variables, usually by some fairly simple mathematical relationship. For example with linear data, as distinct from circular data, a very common model for the regression of the mean value of Y on a variable X is a straight line relationship

$$E(Y|X = x) = a + bx \tag{6.1}$$

(where $E(Y|X = x)$ should be read as 'the mean value of Y given that X takes the value x'). If a model can be found which explains the systematic

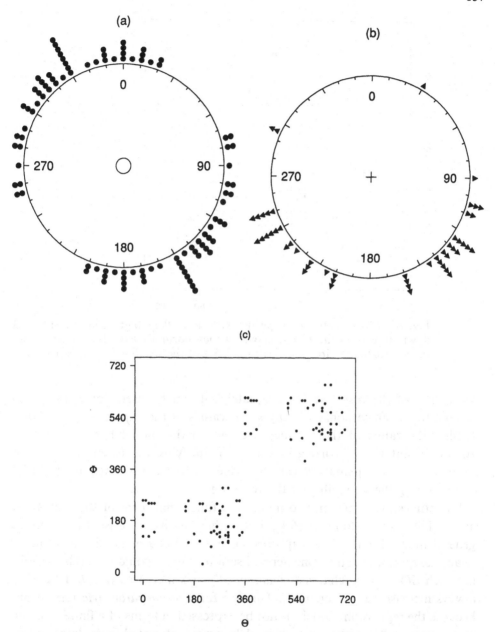

Fig. 6.2 Orientations of the nests of 50 noisy scrub birds along the bank of a creek bed, together with the corresponding directions of creek flow at the nearest point to the nest. (a) Nest orientations. (b) Directions of creek flow. (c) Joint data plot (after converting nest orientations to vectors). See Example 6.2.

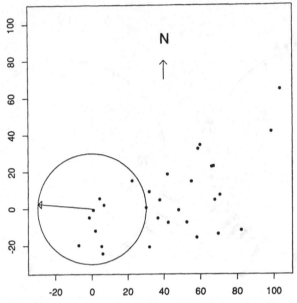

Fig. 6.3 Movements of 31 periwinkles after they had been transplanted downshore from the height at which they normally live. The arrow shows the approximate direction (bearing 275°) of the sea. See Example 6.3.

variability of the response Y 'adequately', it can be used, for example, to predict the mean response for any given values of the explanatory variables (within the range of their values over which the model has been fitted). An important use of correlations is in quantifying the extent to which a proposed set of explanatory variables does indeed contribute 'significantly' to explaining the variability of the response.

Sometimes, we prefer not to model the functional form of the regression as explicitly as is done in (6.1), but rather, to hypothesise that a more general model holds which captures some of the characteristics of a linear relationship. For example, one generalisation of (6.1) would be a relationship in which $E(Y|X = x)$ changed *monotonically* as x increased (e.g. $E(Y|X = x)$ always increased as x increased). This is a form of *nonparametric* regression, because the regression model cannot be expressed in terms of a finite number of parameters. Nevertheless such models can be estimated from data, and in the exploratory phase of data analysis the fitted model can suggest simpler and better models to be fitted. There are measures of general association which quantify some of these relationships as well.

Sections 6.2 and 6.3 treat separately the cases of association between a circular random variable Θ and linear random variable X, and between a circular random variable Θ and another circular random variable Φ. In each case, we begin with measures of general association (which are based on few assumptions) and then proceed to the respective circular–linear or circular–circular models corresponding to (6.1). Section 6.3.4 gives an *omnibus* test for independence of two circular random variables when no particular form of dependence is being considered as an alternative model. Section 6.4 then describes a class of regression models for the dependence of the mean direction or dispersion of a circular random variate on one or more explanatory variables (or *covariates*). For most of **§6.4** it is assumed that the covariates are linear; Section 6.4.5 contains a few brief remarks about the situation when the covariates themselves may be circular.

6.2 Linear–circular association and circular–linear association

6.2.1 Introduction

Before attempting to measure the association between an angular random variable Θ and a linear random variable X, it is crucial to look one step ahead in the data analysis: is the purpose to fit a regression model to

- predict the mean value of X given $\Theta = \theta$ (*linear–circular association*), or
- to fit a model to predict the mean value of Θ given $X = x$ (*circular–linear association*)?

The regression models for these two cases are quite different, and their associated measures of association differ also.

In the linear–circular case, we focus on measuring association between X and Θ with a (possible) view to predicting the mean value of X for a given value θ of Θ. A simple regression model for this type of association has the form

$$E(X|\Theta = \theta) = a_0 + b_0 \cos(\theta - \theta_0) \tag{6.2}$$

which can be rewritten as

$$E(X|\Theta = \theta) = a_0 + a \cos\theta + b \sin\theta$$

In the latter form, it is a simple linear regression model (linear in the regression variables $\cos\theta$ and $\sin\theta$, that is) and can be fitted routinely by methods in any general statistical package, so we shall not discuss the regression aspect further. However, it is useful to have the model in mind in motivating a discussion of association between X and Θ. In §6.2.2, we describe a more general version of this model, together with methods for

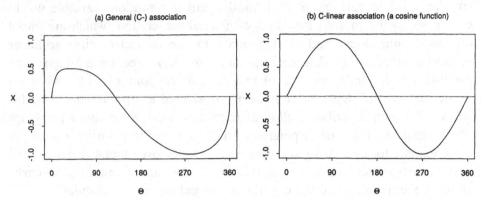

Fig. 6.4 *C*-association and *C*-linear association between a linear random variable X and a circular random variable Θ. (a) *C*-association. (b) *C*-linear association (a cosine function).

assessing the degree to which Θ and X are related by the general model. In §**6.2.3**, this model is studied directly.

Measuring circular–linear association with a view to predicting the mean direction of Θ for a given value of X presents rather different problems. It turns out that this problem can be transformed into one of measuring the association between Θ and another angular random variable; this is described in Example 6.9.

6.2.2 Linear–circular association: C-association

The relationship (6.2) can be thought of as a curve on the surface of a cylinder: as θ performs a cycle from 0 to 360°, $E(X|\Theta = \theta)$ performs a 'sine-wave' oscillation over the range. We term this *C-linear association*, and discuss it in §**6.2.3**. A more general form of (6.2) is any curve which has one maximum and one minimum over the range $(0, 2\pi)$, and whose values match at $\theta = 0$ and $\theta = 2\pi$, that is, a periodic function. An example is shown in Figure 6.4(b). We call this a model of *C-association*. Two methods are available for assessing the extent of *C*-association between Θ and X. One method is based on estimating a parameter λ which quantifies the amount of *C*-association directly; the estimate can then be tested to see if it differs significantly from zero, in which event a confidence interval for λ can be calculated. The other method is far simpler to implement, but just yields a test of the hypothesis $\lambda = 0$, so we discuss it first.

Given a random sample of data $(\theta_1, x_1), \ldots, (\theta_n, x_n)$, let $r_1, \ldots, r_n$ and $s_1, \ldots, s_n$ be the respective sets of circular ranks (§**2.3.3**) of $\theta_1, \ldots, \theta_n$ and ordinary ranks of $x_1, \ldots, x_n$. Calculate

$$D_n = a_n(T_c^2 + T_s^2) \tag{6.3}$$

where

$$T_c = \sum_{i=1}^{n} s_i \cos r_i, \quad T_s = \sum_{i=1}^{n} s_i \sin r_i \tag{6.4}$$

and

$$a_n = \begin{cases} 1/[1 + 5\cot^2(\pi/n) + 4\cot^4(\pi/n)] & n \text{ even.} \\ 2\sin^4(\pi/n)/[1 + \cos(\pi/n)]^3 & n \text{ odd.} \end{cases} \tag{6.5}$$

D_n is a quantity which lies between 0 and 1, with values near zero suggesting little evidence for C-association, and values close to 1 indicating a marked degree of dependence of this form. To assess the significance or otherwise of D_n, calculate

$$U_n = 24(T_c^2 + T_s^2)/(n^3 + n^2) \tag{6.6}$$

The hypothesis that X and Θ are independent will be rejected in favour of the alternative of C-association if U_n is too large.

Critical values. For $n \leq 100$, compare U_n with the appropriate (interpolated) significance point in Appendix A10. For $n > 100$, the significance probability of U_n is $\exp(-U_n^2/2)$; alternatively, for a test at level $100\alpha\%$, reject the hypothesis if $U_n \geq -2\log\alpha$. Note that for $n \leq 100$, we could calculate the significance probability of D_n using a permutation test, as in §**8.5**.

> **Example 6.4** For the data on wind direction and ozone level described in Example 6.1, it seems reasonable to suppose that one might wish to predict the mean level of ozone for a given wind direction, assuming that there is some dependence of X on Θ. The value of D_n from (6.4) is 0.502. To assess this, we calculate the associated quantity $U_n = 6.73$, which just exceed the tabulated 5% point of the null distribution in Appendix A10, so there is some evidence of association between X and Θ.

References and footnotes. The test is due to Mardia (1976).

We now turn to a method which allows us to estimate the extent of C-association. Let $P_1 = (\theta_1, x_1), \ldots, P_4 = (\theta_4, x_4)$ be four data points arranged in cyclic order according to their θ-values, with

$$0 < \theta_1 < \theta_2 < \theta_3 < \theta_4 \leq 2\pi \tag{6.7}$$

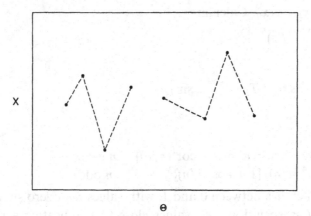

X

Θ

Fig. 6.5 *C-concordant configurations of (X, Θ), for assessing C-association.*

We call the points P_1, P_2, P_3, P_4 *C-concordant* if their x-values have either of the patterns shown in Figure 6.5. To express this in symbols, set

$$d_1 = \text{sign}(x_1 - x_2), \ d_2 = \text{sign}(x_2 - x_3), \ d_3 = \text{sign}(x_3 - x_4), \ d_4$$
$$= \text{sign}(x_4 - x_1) \tag{6.8}$$

Then P_1, P_2, P_3, P_4 are *C-concordant* if

$$(d_1, d_2, d_3, d_4) = (-1, 1, -1, 1) \text{ or } (1, -1, 1, -1) \tag{6.9}$$

and *C-discordant* $(d_1, d_2, d_3, d_4) =$ any of the other 12 possible four–sequences of -1 and 1:

$$(-1, -1, -1, 1), \ (-1, 1, -1, -1), \ \ldots, \ (-1, 1, 1, 1)$$

(The two sequences $(-1, -1, -1, -1)$ and $(1, 1, 1, 1)$ are impossible.)
 Define

$$\psi(P_1, P_2, P_3, P_4) = \begin{cases} 1 & (P_1, P_2, P_3, P_4) \ \textit{C-concordant} \\ 0 & (P_1, P_2, P_3, P_4) \ \textit{C-discordant} \end{cases} \tag{6.10}$$

The parameter we wish to estimate is

$\lambda =$ Probability that four random pairs $(\Theta_1, X_1), \ldots, (\Theta_4, X_4)$

are *C-concordant* $\tag{6.11}$

Given a random sample of data $P_1 = (\theta_1, x_1), \ldots, P_n = (\theta_n, x_n)$, the estimate of λ is

$$\hat{\lambda}_n = \sum_{\substack{\textit{all distinct four-subsets} \\ i,j,k,l \ \textit{of} \ (1,\ldots,n)}} \psi(P_i, P_j, P_k, P_l) / \binom{n}{4} \tag{6.12}$$

When Θ and X are independent, $\lambda = \frac{2}{3}$. Larger values of the sample estimate $\widehat{\lambda}_n$ suggest some degree of C-monotone relationship between Θ and X, and smaller values an ordinary monotone relationship (as between two linear random variables). The maximum possible value of $\widehat{\lambda}_n$ is unknown (except for very small sample sizes), but is less than 1; the minimum possible value is 0.

Computational notes

1. Treatment of ties between θ or between x. Omit the corresponding ψ value if (P_i, P_j, P_k, P_l) has ties between θ or x, and reduce $\binom{n}{4}$ in the denominator of (6.12) by the number of ψ-values so omitted.

2. Calculation of ψ-values. This can be effected rapidly, because the function ψ (cf. (6.10)) depends only on the signs of $x_i - x_j, x_j - x_k, x_k - x_l, x_l - x_i$. Suppose that the data set $(\theta_1, x_1), \ldots, (\theta_n, x_n)$ has been arranged so that the θ are in ascending (or just cyclic) order. Set up an integer vector L of length 14, with $L(5) = L(10) = 1$, all other L-values 0. Also, if storage space is not a problem, set up the $n \times n$ integer matrix $\mathbf{K}$ with elements K_{ij}, where

$$K_{ij} = \begin{cases} 1 & x_i > x_j \\ 0 & \text{otherwise} \end{cases} \tag{6.13}$$

Then

$$\psi(P_i, P_j, P_k, P_l) = L(m) \tag{6.14}$$

where

$$m = 8K_{ij} + 4K_{jk} + 2K_{kl} + K_{li} \tag{6.15}$$

or, if n is too large to set up $\mathbf{K}$,

$$m = 8u(x_i - x_j) + 4u(x_j - x_k) + 2u(x_k - x_l) + u(x_l - x_i) \tag{6.16}$$

with

$$u(x) = \begin{cases} 1 & x > 0 \\ 0 & \text{otherwise} \end{cases} \tag{6.17}$$

3. Complete and incomplete statistics. For $n \leq 20$, it is feasible to calculate $\widehat{\lambda}_n$ as defined in (6.12); an algorithm such as NEXKSB in Nijenhuis & Wilf (1978) can be used to enumerate all the four-subsets (i, j, k, l) of $(1, \ldots, n)$. For $n > 20$, the complete enumeration of $\binom{n}{4}$ subsets may not be possible. In this case, calculate the *incomplete* statistic

$$\widehat{\lambda}_{n,M} = \sum_{\substack{M \text{ randomly selected four-subsets} \\ i,j,k,l \text{ of } (1,\ldots,n)}} \psi(P_i, P_j, P_k, P_l)/M \tag{6.18}$$

for a large number M (e.g. using the algorithm RANKSB in Nijenhuis & Wilf, 1978).

To test the hypothesis H_0: $\lambda = \frac{2}{3}$ against the alternative H_1: $\lambda > \frac{2}{3}$, calculate $\widehat{\lambda}_n$ ($n \leq 20$) or $\widehat{\lambda}_{n,M}$ ($n > 20$). The hypothesis of no C-association between Θ and X is rejected if $\widehat{\lambda}_n(\widehat{\lambda}_{n,M})$ is too large.

Critical values.

(a) $n \leq 8$. Some cumulative probabilities $\mathrm{Prob}\,(\widehat{\lambda}_n \leq x$, assuming H_0 true) are given in Appendix A11(a).

(b) $9 \leq n \leq 20$. Use a randomisation test (cf. §**8.5**).

(c) $n > 20$.   If it is feasible to calculate the complete statistic $\widehat{\lambda}_n$, refer $\widehat{\Lambda} = n(\widehat{\lambda}_n - \text{case } \frac{2}{3})$ to the critical values in A11(c). If the statistic $\widehat{\lambda}_{n,M}$ is used, for a test at level α calculate the modified value $L(\alpha)$ given in Appendix A11(b). The hypothesis that $\lambda = \frac{2}{3}$ (no C-association) is rejected at the $100\alpha\%$ level if $L(\alpha) > 0$. Alternatively, use a permutation test to estimate the significance probability, as in §**8.5**. A detailed discussion of how the large-sample distribution relates to the value of the ratio n^2/M is given in Fisher & Lee (1981, §3.2). Generally, if $\widehat{\lambda}_{n,M}$ is calculated, a randomisation test is to be preferred.

> **Example 6.5** For the data on wind direction and ozone level described in Example 6.1 and analysed in Example 6.4, we get $\widehat{\lambda}_{19} = 0.7327$ for the complete statistic. Here, the sample size is just below 20; however, for comparative purposes, we shall assess significance using both the asymptotic distribution and an incomplete version of $\widehat{\lambda}_n$. We get $\widehat{\Lambda} = 1.255$, which lies between the $2\frac{1}{2}\%$ and 1% point in Appendix A11(c), indicating some evidence against the hypothesis H_0: $\lambda = \frac{2}{3}$ i.e. against the independence of X and Θ in favour of an alternative of C-association (as was found in Example 6.4). Using the incomplete statistic $\widehat{\lambda}_{19,M}$ based on, say, $M = 1000$ random subsets, we get $\widehat{\lambda}_{19,M} = 0.707$, which lies between the 10% and 5% critical values. A more accurate assessment of this statistic would be obtained by using a randomisation test.

References and footnotes. The statistic $\widehat{\lambda}_n$, and a more complicated one based on five data pairs P_1, P_2, P_3, P_4, P_5, are due to Fisher & Lee (1981). Note that Table 3 of that paper contains errors: the hypotheses being tested are $\pi_4 = \frac{2}{3}$ ($\lambda = \frac{2}{3}$, here) and $\pi_5 = \frac{1}{3}$. The remarks made above, relating to the minimum and maximum values of $\widehat{\lambda}_n$, correct the corresponding statements in that paper.

6.2.3 Linear–circular association: C-linear association

A simple way to assess the extent of C-linear dependence of $E(X|\Theta = \theta)$ on θ (cf. (6.2)) is to compute the estimate of multiple correlation of X with $(\cos\Theta, \sin\Theta)$. For any set of numbers $(u_1, v_1), \ldots, (u_n, v_n)$, let $r\{(u_1, v_1), \ldots, (u_n, v_n)\}$ denote the sample correlation

$$\frac{\sum\limits_{i=1}^{n} (u_i - \bar{u})(v_i - \bar{v})}{\left[\sum\limits_{i=1}^{n} (u_i - \bar{u})^2 \sum\limits_{i=1}^{n} (v_1 - \bar{v})^2\right]^{\frac{1}{2}}}, \quad \bar{u} = \sum\limits_{j=1}^{n} u_j/n, \ \bar{v} = \sum\limits_{j=1}^{n} v_j/n \tag{6.19}$$

Calculate

$$\left.\begin{aligned}
r_{12} &= r\{(x_1, \cos\theta_1), \ldots, (x_n, \cos\theta_n)\} \\
r_{13} &= r\{(x_1, \sin\theta_1), \ldots, (x_n, \sin\theta_n)\} \\
r_{23} &= r\{(\cos\theta_1, \sin\theta_1), \ldots, (\cos\theta_n, \sin\theta_n)\}
\end{aligned}\right\} \tag{6.20}$$

and then

$$R_n^2 = (r_{12}^2 + r_{13}^2 - 2r_{12}r_{13}r_{23})/(1 - r_{23}^2) \tag{6.21}$$

The hypothesis of no C-linear association is rejected if R_n^2 is too large.

Critical values. Use a randomisation test as described in §**8.5**. As mentioned in §**6.2.1**, if there is evidence of C-linear association, the associated regression model can now be fitted by standard methods.

References and footnotes. The test is due to Mardia (1976); it was further developed by Liddell & Ord (1978), who also found the exact distribution.

6.3 Circular–circular association

6.3.1 Introduction

We begin, in §**6.3.2**, with a general model for association (called *T-monotone association*) between Θ and Φ, which corresponds to *monotone* association of *real* variables as measured by Kendall's tau or Spearman's rho. Then in §**6.3.3** we describe a form of association (called *T-linear association*) corresponding to *linear* association of real variables, and a suitable correlation coefficient. Finally, a totally general test of independence is described in §**6.3.4**, which can be used in situations where it is desired to detect *any* type of association (T-linear, T-monotone or otherwise) between Θ and Φ.

(a) C-concordance (positive association)   (b) C-discordance (negative association)

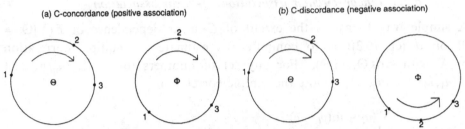

Fig. 6.6 Basic forms of positive and negative association (T-monotone association) of two circular random variables, using three pairs of measurements. (a) Positive association: the triple of pairs is concordant. (b) Negative association: the triple of pairs is discordant.

6.3.2 Circular–circular association: T-monotone association

Let $p_1 = (\theta_1, \phi_1), \ldots, p_n = (\theta_n, \phi_n)$ be a sample of independent measurements of a pair of circular random variables $P = (\Theta, \Phi)$. A basic type of relationship between Θ and Φ, as manifested in the sample, is depicted in Figure 6.6 using $(\theta_1, \phi_1), (\theta_2, \phi_2), (\theta_3, \phi_3)$, where we have assumed that θ_1, θ_2 and θ_3 plot in a clockwise direction. The corresponding values ϕ_1, ϕ_2, ϕ_3 can plot either in a clockwise (Figure 6.6(a)) or a counterclockwise (Figure 6.6(b)) direction; we term these configurations *T-concordant* and *T-discordant* respectively (the 'T' deriving from the fact that the pairs (θ_i, ϕ_i) actually lie on the surface of a torus, or doughnut), and the form of association *T-monotone* association. *T-concordant* triples suggest *positive* association between Θ and Φ and *T-discordant* triples *negative* association. This leads to a simple measure of association between Θ and Φ: if P_1, P_2 and P_3 are three random pairs with the same distribution as P, define

$$\Delta = \text{Prob}\{(P_1, P_2, P_3) \text{ are } T\text{-concordant}\}$$
$$- \text{Prob}\{(P_1, P_2, P_3) \text{ are } T\text{-discordant}\} \tag{6.22}$$

The values $\Delta = 1$ and $\Delta = -1$ correspond to Θ and Φ being completely dependent, e.g. $\Theta = \Phi$ or $\Theta = -\Phi$. However, $\Delta = 0$ does *not* signify independence, merely that Θ and Φ do not have any association of this form.

There are two statistics we can use to assess T-monotone association. One gives a direct estimate of Δ, which can be thus used to obtain a confidence interval for Δ, the other assesses the hypothesis $\Delta = 0$ a bit more indirectly but is computationally faster.

(a) The correlation coefficient $\widehat{\Delta}_n$

To estimate Δ directly, define

$$\delta(p_1, p_2, p_3) = \text{sign}\,(\theta_1 - \theta_2)\,\text{sign}\,(\theta_2 - \theta_3)\,\text{sign}\,(\theta_3 - \theta_1)$$
$$\times\,\text{sign}\,(\phi_1 - \phi_2)\,\text{sign}\,(\phi_2 - \phi_3)\,\text{sign}(\phi_3 - \phi_1) \qquad (6.23)$$

so that δ takes the values

$$\begin{cases} 1 & (p_1, p_2, p_3)\ \text{concordant} \\ -1 & (p_1, p_2, p_3)\ \text{discordant} \\ 0 & \text{ties amongst } \theta_1, \theta_2, \theta_3 \text{ or } \phi_1, \phi_2, \phi_3 \end{cases} \qquad (6.24)$$

Then estimate Δ by

$$\widehat{\Delta}_n = \sum_{1 \le i < j < k < n} \delta(p_i, p_j, p_k) / [\tbinom{n}{3} - N_0] \qquad (6.25)$$

where N_0 is the number of zero δ-values. The hypothesis $H_0 : \Delta = 0$ is rejected if $\widehat{\Delta}_n$ differs too much from zero. Note that there are of the order of $n^3/6$ terms to be summed in equation (6.25), which may not be practicable for larger sample sizes. The value of $\widehat{\Delta}$ can then be approximated using a large number M of randomly selected three-subsets. These subsets can generated using the algorithm RANKSB in Nijenhuis & Wilf (1978, p. 39), for example. The statistic calculated is then

$$\widehat{\Delta}_{n,M} = \sum_{\substack{M\ randomly\ selected\ three-subsets \\ i,j,k\ of\ (1,...,n)}} \delta(p_i, p_j, p_k) / (M - N_0) \qquad (6.26)$$

Critical values.

(a) $n < 8$. Appendix A12(a) gives the probability distribution of $\widehat{\Delta}_n$. Suppose the sample value of $\widehat{\Delta}_n$ is $\widehat{\Delta}_0$.

To test H_0 against $H_1 : \Delta \neq 0$, calculate $\text{Prob}\,(|\,\widehat{\Delta}_n\,| \ge |\,\widehat{\Delta}_0\,|)$ by adding the values $\text{Prob}\,(n\widehat{\Delta}_n = x)$ for $x \ge n\widehat{\Delta}_0$, and doubling the total of these probabilities. This gives the significance probability of $|\,\widehat{\Delta}_n\,|$.

To test H_0 against $H_1 : \Delta > 0$ [$\Delta < 0$], calculate $\text{Prob}\,(\widehat{\Delta}_n \ge \widehat{\Delta}_0)$ [$\text{Prob}\,(\widehat{\Delta}_n \le \widehat{\Delta}_0)$] by adding the values $\text{Prob}\,(n\widehat{\Delta}_n = x)$ for $x \ge n\widehat{\Delta}_0$ [$x \le n\widehat{\Delta}_0$]. This gives the significance probability of $\widehat{\Delta}_n$.

(b) $n \ge 8$. Appendix A12(b) contains upper critical values of the distribution of $n\widehat{\Delta}_n$. To test H_0 against $H_1 : \Delta \neq 0$, compare $|\,n\widehat{\Delta}_n\,|$ or $|\,n\widehat{\Delta}_{n,M}\,|$ with $x_{n,\frac{1}{2}\alpha}$ in Appendix A12(b). To test H_0 against $H_1 : \Delta > 0$ [$\Delta < 0$] compare $n\widehat{\Delta}_n$ or $n\widehat{\Delta}_{n,M}$ with $x_{n,\alpha}$ [$-x_{n,\alpha}$].

If H_0 is rejected, we can obtain a large-sample confidence interval for Δ as follows. Calculate

$$\hat{\sigma}_\Delta^2 = \sum_{l=1}^{n} (\hat{\Delta}_n^{(l)} - \hat{\Delta}_n)^2 \qquad (6.27)$$

where

$$\hat{\Delta}_n^{(l)} = \sum_{\substack{1 \le i < j \le n \\ i,j \ne l}} \delta(p_l, p_i, p_j) / \left[\binom{n-1}{2} - N_0 \right] \quad l = 1, \ldots, n \qquad (6.28)$$

(N_0 being the number of zero δ-values).

Then, for $n \ge 25$,

$$\hat{\Delta}_n \pm 3\hat{\sigma}_\Delta z_{\frac{1}{2}\alpha} / \sqrt{n} \qquad (6.29)$$

is an approximate $100(1 - \alpha)\%$ confidence interval for Δ, where $z_{\frac{1}{2}\alpha}$ is the upper $100(\frac{1}{2}\alpha)\%$ point of the $N(0, 1)$ distribution in Appendix A1. For smaller values of n, a bootstrap method (preferably using the iterated method) can be used to calculate an approximate confidence interval.

(b) The correlation coefficient $\hat{\Pi}_n$

An alternative, and computationally faster, method of testing the hypothesis of no T-monotone association is based on the *circular ranks* of the data (cf. §2.3.3). Let $\gamma_1, \ldots, \gamma_n$ be the circular ranks of $\theta_1, \ldots, \theta_n$, and $\epsilon_1, \ldots, \epsilon_n$ the circular ranks of $\phi_1, \ldots, \phi_n$ (cf. (2.35)). Calculate

$$\hat{\Pi}_n = (4/n^2) \sum_{1 \le i < j \le n} \sin(\gamma_i - \gamma_j) \sin(\epsilon_i - \epsilon_j) \qquad (6.30)$$

The hypothesis that Θ and Φ are independent is rejected if $\hat{\Pi}_n$ differs too much from zero. The alternative models of positive or negative association we have in mind are those described in Figure 6.6. Equation (6.30) has been written in this form to emphasise its relationship to the statistic $\hat{\rho}_T$ in §6.3.3 below. However, like $\hat{\rho}_T$ it has an alternative representation which is preferable for computational purposes:

$$\hat{\Pi}_n = (4/n^2)(AB - CD) \qquad (6.31)$$

where

$$\left. \begin{array}{ll} A = \sum_{i=1}^{n} \cos \gamma_i \cos \epsilon_i, & B = \sum_{i=1}^{n} \sin \gamma_i \sin \epsilon_i \\[2mm] C = \sum_{i=1}^{n} \cos \gamma_i \sin \epsilon_i, & D = \sum_{i=1}^{n} \sin \gamma_i \cos \epsilon_i \end{array} \right\} \qquad (6.32)$$

Critical values. The null and alternative hypotheses of interest are exactly as for $\widehat{\Delta}_n$ above. The testing procedures are also the same, except that the statistics $(n-1)\widehat{\Pi}_n$ and $\mid (n-1)\widehat{\Pi}_n \mid$ are referred to corresponding tables in Appendix A13.

Example 6.6 For the noisy scrub bird data of Example 6.2, we consider testing the hypothesis of no association between nest orientation and creek orientation. Here we are faced with the added difficulty that creek orientation (as determined by the direction of current flow) is vectorial whereas nest orientation is axial. However, for rank correlations, this is immaterial, because converting the nest orientations to vectors would not change their cyclic ordering and the rank correlations depend on the data *only* through their cyclic orders. It is not easy to see any relationship between the variables in the plot, so we shall calculate the two measures of association described above. For sample size $n = 50$, it is not practicable to calculate the complete statistic $\widehat{\Delta}_n$ in (6.25), so the incomplete version in (6.26) was computed with $M = 10\,000$. There are numerous ties in the data; however, the sample size is probably large enough to make the assessment of the significance of the statistics reasonable. We get

$$\widehat{\Delta}_{n,M} = -0.0209, \; n\widehat{\Delta}_{n,M} = -1.05, \; \widehat{\Pi}_n = -0.0213, \; (n-1)\widehat{\Pi}_n$$
$$= -1.04$$

On referring the values of $\widehat{\Delta}_{n,M}$ and $(n-1)\widehat{\Pi}_n$ to their respective tables of critical values, we find that neither indicates any evidence for dependence of nest orientation on direction of current flow. (*Note*: there are a lot of ties in both the nest orientation and the creek direction data, leading to a value of N_0 in (6.26) of about 3000. However, the value of $n\widehat{\Delta}_{n,M}$ is so small that this is unlikely to have affected the conclusion we drew in any material way. Had the sample size been smaller, we could have used a randomisation method, as described in §**8.5**, to evaluate the significance of the statistics.)

Example 6.7 Figure 6.7 shows measurements of wind directions at 6.00am and 12.00 noon, on each of 21 consecutive days, at a weather station in Milwaukee. (The data are listed in Appendix B21.) We consider testing the assumption of no relationship between the 6.00am and 12.00 noon readings, against the alternative that they are positively associated. For these data, we get

$$\widehat{\Delta}_n = 0.214, \; n\widehat{\Delta}_n = 4.49, \; \widehat{\Pi}_n = 0.227, \; (n-1)\widehat{\Pi}_n = 4.53$$

Both $n\widehat{\Delta}_n$ and $(n-1)\widehat{\Pi}_n$ exceed the 1% points of their respective null distributions, providing strong evidence for positive association of pairs of measurements made on the same day.

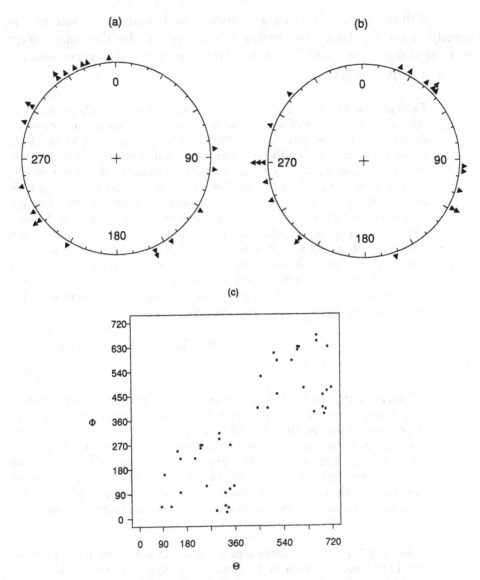

Fig. 6.7 Measurements of wind direction at 6.00am and 12 noon on 21 consecutive days, at a weather station in Milwaukee. (a) 6.00am measurements (b) 12 noon measurements (c) joint circular data plot. The marginal sample values in (a) and (b) have the appearance of uniformity. There is some suggestion from the graph that the 6.00am and 12 noon measurements are associated. See Example 6.7.

References and footnotes. The statistic $\widehat{\Delta}_n$ is due to Fisher & Lee (1982); the statistic $\widehat{\Pi}_n$ is due to Fisher & Lee (1982, 1983) and is based on an unsigned statistic proposed by Mardia (1975).

6.3.3 Circular–circular association: T-linear association

A simple analogue of linear association between real variables is given by the following model:

$$\Phi = \Theta + \theta_0 \text{ (modulo 360°)} \tag{6.33}$$

or

$$\Phi = -\Theta + \theta_0 \text{ (modulo 360°)} \tag{6.34}$$

This is termed T-linear association, with (6.33) describing complete positive association and (6.34) complete negative association. (Note that it corresponds to a special case of linear association, as there is no scale parameter: a more general form $\Phi = a\Theta + \theta_0$ for some scale parameter a turns out to be difficult to interpret, let alone estimate statistically.) The extent of such association can be estimated from the sample using the statistic

$$\widehat{\rho}_T = \frac{\displaystyle\sum_{1\le i<j\le n} \sin(\theta_i - \theta_j)\sin(\phi_i - \phi_j)}{\left[\displaystyle\sum_{1\le i<j\le n}\sin^2(\theta_i-\theta_j)\sum_{1\le i<j\le n}\sin^2(\phi_i-\phi_j)\right]^{\frac{1}{2}}} \tag{6.35}$$

For computational purposes, the alternative form

$$\widehat{\rho}_T = 4(AB - CD)/\left[(n^2 - E^2 - F^2)(n^2 - G^2 - H^2)\right]^{\frac{1}{2}} \tag{6.36}$$

where

$$\left.\begin{aligned}
A &= \sum_{i=1}^{n}\cos\theta_i\cos\phi_i, & B &= \sum_{i=1}^{n}\sin\theta_i\sin\phi_i \\
C &= \sum_{i=1}^{n}\cos\theta_i\sin\phi_i, & D &= \sum_{i=1}^{n}\sin\theta_i\cos\phi_i \\
E &= \sum_{i=1}^{n}\cos(2\theta_i), & F &= \sum_{i=1}^{n}\sin(2\theta_i) \\
G &= \sum_{i=1}^{n}\cos(2\phi_i), & H &= \sum_{i=1}^{n}\sin(2\phi_i)
\end{aligned}\right\} \tag{6.37}$$

may be used. The closer $\widehat{\rho}_T$ is to 1, or to −1, the greater the extent to which (6.33), or (6.34), explains the relationship between Θ and Φ. The hypothesis of no T-linear association is rejected if $\widehat{\rho}_T$ differs too much from 0.

Critical values. For $n < 25$, use a randomisation test (cf. §8.5) based on the quantity $T = AB - CD$ (cf. (6.34)). For $n \geq 25$, the test depends on the marginal distributions of Θ and Φ:

(i). If either distribution has mean resultant length 0 (cf. §4.3), for example if the distributions are uniform, the statistic $n\hat{\rho}_T$ has, approximately, a double-exponential distribution with density $\frac{1}{2}e^{-|x|}$. For a test at level $100\alpha\%$,

$$\text{reject } H_0 : \rho_T = 0 \text{ in favour of } H_1 : \rho_T \neq 0 \text{ if } | n\hat{\rho}_T | > -\log \alpha;$$
$$\text{reject } H_0 : \rho_T = 0 \text{ in favour of } H_1 : \rho_T > 0 \, [\rho_T < 0]$$
$$\text{if } n\hat{\rho}_T > -\log(2\alpha) \, [n\hat{\rho}_T < \log 2\alpha].$$

(ii). If the mean resultant lengths of the distributions can be assumed to be non-zero, calculate the separate sample mean directions $\bar{\theta}$ and $\bar{\phi}$ and mean resultant lengths $\bar{R}_\theta$ and $\bar{R}_\phi$, the second central moments

$$\left. \begin{array}{ll} \hat{\alpha}_2(\theta) = (1/n) \sum_{i=1}^{n} \cos 2(\theta_i - \bar{\theta}), & \hat{\beta}_2(\theta) = (1/n) \sum_{i=1}^{n} \sin 2(\theta_i - \bar{\theta}) \\ \hat{\alpha}_2(\phi) = (1/n) \sum_{i=1}^{n} \cos 2(\phi_i - \bar{\phi}), & \hat{\beta}_2(\phi) = (1/n) \sum_{i=1}^{n} \sin 2(\phi_i - \bar{\phi}) \end{array} \right\}$$

$$(6.38)$$

and then the quantities

$$\left. \begin{array}{l} U(\theta) = (1 - \hat{\alpha}_2^2(\theta) - \hat{\beta}_2^2(\theta))/2 \\ U(\phi) = (1 - \hat{\alpha}_2^2(\phi) - \hat{\beta}_2^2(\phi))/2 \\ V(\theta) = \bar{R}_\theta^2 (1 - \hat{\alpha}_2(\theta)) \\ V(\phi) = \bar{R}_\phi^2 (1 - \hat{\alpha}_2(\phi)) \end{array} \right\}$$

$$(6.39)$$

Finally, set

$$Z = n^{\frac{1}{2}} U(\theta) U(\phi) \hat{\rho}_T / (V(\theta) V(\phi))^{\frac{1}{2}} \tag{6.40}$$

Let $z_{\frac{1}{2}\alpha}$ and z_α denote the upper $100(\frac{1}{2}\alpha)\%$ and $100\alpha\%$ points of the $N(0,1)$ distribution in Appendix A1. Then $H_0: \rho_T = 0$ is rejected in favour of $H_1: \rho_T \neq 0$ if $| Z | > z_{\frac{1}{2}\alpha}$; and $H_0: \rho_T = 0$ is rejected in favour of $H_1: \rho_T > 0 \, [\rho_T < 0]$ if $Z > z_\alpha \, [Z < -z_\alpha]$.

If we decide that $\rho_T \neq 0$ then, for $n \geq 25$, an approximate 95% confidence interval for ρ_T can be computed. Let $\hat{\rho}_{-i}$ be the estimate of ρ_T calculated from the sample with (θ_i, ϕ_i) omitted, and let $\hat{\rho}_T^{(i)} = n\hat{\rho}_T - (n-1)\hat{\rho}_{-i}, i =$

$1, \ldots, n$. The jackknife estimate of ρ_T is

$$\widehat{\rho}_{T,J} = \sum_{i=1}^{n} \widehat{\rho}_T^{(i)}/n \tag{6.41}$$

with jackknife estimate of variance

$$s^2 = \sum_{i=1}^{n} (\widehat{\rho}_T^{(i)} - \widehat{\rho}_{T,J})^2/(n-1) \tag{6.42}$$

An approximate $100(1-\alpha)\%$ confidence interval for ρ_T is then

$$\widehat{\rho}_{T,J} \pm n^{-\frac{1}{2}} s z_{\frac{1}{2}\alpha} \tag{6.43}$$

where $z_{\frac{1}{2}\alpha}$ is the upper $100(\frac{1}{2}\alpha)\%$ point of the $N(0,1)$ distribution in Appendix A1. For smaller sample sizes ($n < 25$) use a bootstrap method (§8.4).

> **Example 6.8** Consider the data on wind directions at 6.00am and 12.00 noon analysed in Example 6.7, with the same null and alternative hypotheses. We get $\widehat{\rho}_T = 0.191$, corresponding to a significance probability of about 0.01, based on 1000 random permutations of $\phi_1, \ldots, \phi_{21}$ compared with $\theta_1, \ldots, \theta_{21}$, so there is strong evidence for positive association between the 6.00am and 12 noon readings, in agreement with the analysis in Example 6.7. To calculate a confidence interval for the strength of correlation, we shall use a bootstrap method. In practice, it can be difficult to get bootstrap confidence intervals for correlation coefficients which are close to the desired nominal coverage (e.g. close to 95%), so we shall use the iterated bootstrap method to calibrate the interval. The calibration curve is shown in Figure 6.8; since it deviates somewhat from the 45° line, calibration will make a difference. Suppose we want a 90% confidence interval for ρ_T. Let
>
> $$\widehat{\rho}_{(1)}^* \leq \cdots \leq \widehat{\rho}_{(200)}^*$$
>
> denote the ordered bootstrap estimates of ρ_T, based on $B = 200$ resamples. Using the general method described in §8.4, adapted to correlations, the uncalibrated 90% confidence interval for ρ_T would be $(\widehat{\rho}_{(11)}^*, \widehat{\rho}_{(190)}^*) = (-0.02, 0.41)$, whereas the calibrated interval is $(\widehat{\rho}_{(16)}^*, \widehat{\rho}_{(177)}^*) = (0.01, 0.33)$.

References and footnotes. Many statistics have been suggested for measuring T-linear association: see Jupp & Mardia (1989) for a survey of the literature. The statistic $\widehat{\rho}_T$ is due to Fisher & Lee (1983, 1986).

6.3.4 *A general test for circular–circular association*

In some situations, we may simply wish to carry out a general test of the hypothesis that Θ and Φ are independent. Re-arrange the data points to get

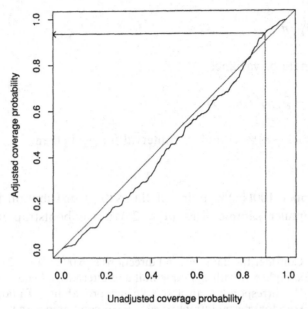

Fig. 6.8 Calibration plot to adjust the size of the approximate 90% bootstrap confidence interval for the correlation between wind directions for the data in Figure 6.7. See Example 6.8.

$(\theta_1^*, \phi_1^*), \ldots, (\theta_n^*, \phi_n^*)$, where $\theta_1^*, \ldots, \theta_n^*$ are in cyclic order, and let $S_1, \ldots, S_n$ be the ranks (*not* circular ranks) of the corresponding ϕ^*-values. Calculate

$$\gamma_n^2 = (1/n^4) \sum_{i=1}^{n} \sum_{j=1}^{n} \left(T_{i,i} + T_{j,j} - T_{i,j} - T_{j,i} \right)^2 \tag{6.44}$$

where

$$T_{ij} = n \min(i, S_j) - i \times S_j, \quad i, j = 1, \ldots, n \tag{6.45}$$

The hypothesis that Θ and Φ are independent is rejected if γ_n^2 is too large.

Critical values. Use a randomisation test (cf. §8.5) based on permuting the ϕ^*-values relative to the θ^*-values (kept fixed). For a test at level $100\alpha\%$, reject $H_0 : \Theta$ and Φ are independent if γ_n^2 falls amongst the upper $100\alpha\%$ of randomisation values.

References and footnotes. The test is due to Rothman (1971), who also gave an asymptotic distribution for γ_n^2. However, the asymptotic distribution is yet to be tabulated, and the sample size beyond which its use is appropriate is unknown. Another test is due to Marriott (1969), who proposed modelling the angular difference between Θ and Φ by a simple one-parameter distribution and testing the hypothesis that the parameter is zero.

6.4 Regression models for a circular response variable

6.4.1 Introduction

Our interest in this section will be in endeavouring to model the variation of a mean direction of a circular response variable Θ in terms of one or more explanatory variables, or *covariates*. For most of the section, the covariates will be linear, and we denote them by $X_1, \ldots, X_k$. (The X_i may be random variables themselves, or may simply be so-called design points, such as a set of equally-spaced values at which the response Θ has been measured.) A common feature of data sets of this type, such as that described in Example 6.3, is that not only is the mean direction of Θ, given particular values $X_1 = x_1, \ldots, X_k = x_k$ of the explanatory variables, dependent on $x_1, \ldots, x_k$, but so is the dispersion of Θ. So in practice, it may be necessary to fit a regression model for both the mean direction and the dispersion of Θ, given $x_1, \ldots, x_k$. Since 'dispersion' is not, in general, a well-defined concept for circular random variables, we shall give methods developed in the context of the von Mises distribution, although the methods relating purely to the mean regression are applicable more generally. Also, the methods will be presented for the case of several explanatory variables $X_1, \ldots, X_k$, with specialisation to the case of a single explanatory variable ($k = 1$).

Thus, we suppose that $(\theta_1, \mathbf{x}_1), \ldots, (\theta_n, \mathbf{x}_n)$ is a set of independent observations, with θ_i being drawn from a von Mises $VM(\mu_i, \kappa_i)$ distribution, $i = 1, \ldots, n$, and where $\mathbf{x}_i$ denotes the vector of explanatory values $(x_{i,1}, \ldots, x_{i,k})'$ associated with θ_i, $i = 1, \ldots, n$. If there is no dependence of *dispersion* on $\mathbf{X}$, then the model will be $VM(\mu_i, \kappa)$, and correspondingly, if there is no dependence of the *mean direction* on $\mathbf{X}$, the model will be $VM(\mu, \kappa_i)$. These two cases are analysed in **§6.4.2** and **§6.4.3** respectively, and the full *mixed* model in **§6.4.4**. Generally, the discussion in these subsections follows closely the description in Fisher & Lee (1992), which can be consulted for further detail. A few comments on the case when a covariate is itself circular are made in **§6.4.5**. Exploratory methods for studying the dependence of an angular variate on one or two linear predictors can be found in Chapter 7 (particularly **§7.2.2** and **§7.3**).

It is important to clarify the difference between the situation in **§6.2.3** and the present case. There, the response variable was linear, and the explanatory

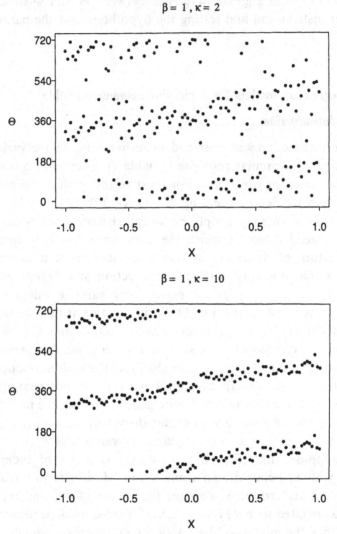

Fig. 6.9 Prototype plots of an angular–linear regression relationship for the
$VM(g(\beta x), \kappa)$ model.

variable circular. If we take the form of the regression model used there
(cf. (6.2)) and attempt to rewrite it as a regression model for the mean
direction of Θ, given $X = x$, we get a 'barber's pole' model, in which the
regression is a curve winding in an infinite number of spirals along the
surface of an infinite cylinder. This is not a satisfactory model to work with
in practice.

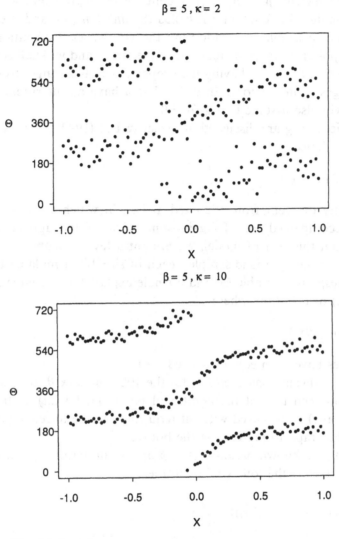

Fig. 6.9 (*cont.*)

6.4.2 Circular regression for the mean direction

Let $(\theta_1, \mathbf{x}_1), \ldots, (\theta_n, \mathbf{x}_n)$ be a set of independent observations, with θ_i being drawn from a von Mises $VM(\mu_i, \kappa)$ distribution, $i = 1, \ldots, n$. We shall suppose that the mean direction μ_i is related to the explanatory variable $\mathbf{X}_i$ by the equation

$$\mu_i = \mu + g(\boldsymbol{\beta}'\mathbf{X}_i) \equiv \mu + g(\beta_1 X_1 + \cdots \beta_k X_k) \tag{6.46}$$

The vector of coefficients $\boldsymbol{\beta} = (\beta_1, \ldots, \beta_k)'$ contains the regression coefficients we seek to estimate. The function g is called the *link function* and is usually selected in advance (similarly to adopting the von Mises distribution as a model). Its purpose is to map the real line to the circle, and we shall consider only monotone link functions having the property that as x ranges from $-\infty$ to ∞, $g(x)$ ranges from $-\pi$ to π. In order that μ have an interpretation as an origin, we will also assume that $g(0) = 0$.

Possible choices for g are discussed in Fisher & Lee (1992). Here, we shall use the specific choice

$$g(u) = 2\tan^{-1}(u) \tag{6.47}$$

(which is slightly different from the cited article, in which an unnecessary factor of $\frac{1}{2}$ accompanied u). To gain some idea of the appearance of data arising from this sort of model, we present a few simulated data sets. Figure 6.9 shows four synthetic samples, each of size 100, simulated from a model with a response variable Θ_i and a single explanatory variable X_i; Θ_i has a $VM(\mu_i, \kappa)$ distribution, where

$$\mu_i = \mu + g(\beta x_i) \tag{6.48}$$

and the x-values have been equally spaced on $(-1, 1)$.

The method of display has been to plot the 100 points (x, θ) in Cartesian coordinates, and then to plot the additional points $(x, \theta + 2\pi)$ so that the relationship (if any) is displayed without requiring the eye to relate patterns at the top of the graph to patterns at the bottom.

To estimate the unknown parameters μ, $\boldsymbol{\beta}$ and κ, an iterative procedure is usually needed. Define the following quantities:

$$u_i = \sin\left(\theta_i - \mu - g(\boldsymbol{\beta}'\mathbf{x}_i)\right) \tag{6.49}$$

$$\mathbf{u}' = (u_1, \ldots, u_n) \tag{6.50}$$

$$\mathbf{X} = \begin{pmatrix} \mathbf{x}'_1 \\ \vdots \\ \mathbf{x}'_n \end{pmatrix} \tag{6.51}$$

$$\mathbf{G} = \mathrm{diag}\left(g'(\boldsymbol{\beta}'\mathbf{x}_1), \ldots, g'(\boldsymbol{\beta}'\mathbf{x}_n)\right) \tag{6.52}$$

$$S = \sum_{i=1}^{n} \sin\left(\theta_i - g(\boldsymbol{\beta}'\mathbf{x}_i)\right)/n \tag{6.53}$$

$$C = \sum_{i=1}^{n} \cos\left(\theta_i - g(\boldsymbol{\beta}' \mathbf{x}_i)\right)/n \qquad (6.54)$$

and

$$R = (S^2 + C^2)^{1/2} \qquad (6.55)$$

Then the *maximum likelihood estimates* (MLE) of the parameters are values which maximise the quantity

$$-n \log I_0(\kappa) + \kappa \sum_{i=1}^{n} \cos\left(\theta_i - \mu - g(\boldsymbol{\beta}' \mathbf{x}_i)\right) \qquad (6.56)$$

and so are solutions to the equations

$$\mathbf{X}'\mathbf{Gu} = 0 \qquad (6.57)$$
$$R \sin \hat{\mu} = S \qquad (6.58)$$
$$R \cos \hat{\mu} = C \qquad (6.59)$$

and

$$A_1(\hat{\kappa}) = R \qquad (6.60)$$

where the function $A_1(\kappa)$ is discussed in §**3.3.6** (see *Note 1*). The following iterative method can be used to solve equations (6.57)–(6.60). Starting with an initial value for $\hat{\boldsymbol{\beta}}$, calculate S, C and R and hence $\hat{\mu}$ and $\hat{\kappa}$ by (6.58)–(6.60). These estimates can then be used to solve (6.57) for an updated version $\hat{\boldsymbol{\beta}}^*$ of $\boldsymbol{\beta}$. Equation (6.57) is itself solved iteratively, using the updating equations

$$\mathbf{X}'\mathbf{G}^2\mathbf{X}(\hat{\boldsymbol{\beta}}^* - \hat{\boldsymbol{\beta}}) = \mathbf{X}'\mathbf{G}^2\mathbf{y} \qquad (6.61)$$

where $\mathbf{y} = (y_1, \ldots, y_n)$ and $y_i = u_i / A_1(\kappa) g'(\hat{\boldsymbol{\beta}}' \mathbf{x}_i)$. We update $\hat{\boldsymbol{\beta}}$ by solving (6.61) and then update $\hat{\mu}$ and $\hat{\kappa}$ by (6.58)–(6.60). This cycle is repeated until convergence.

The large-sample variances and covariances of the regression coefficient estimates $\beta_1, \ldots, \beta_k$ are given by

$$Var\ \hat{\boldsymbol{\beta}} = \frac{1}{\kappa A_1(\kappa)} \left((\mathbf{X}'\mathbf{G}^2\mathbf{X})^{-1} + \frac{(\mathbf{X}'\mathbf{G}^2\mathbf{X})^{-1}\mathbf{X}'\mathbf{gg}'\mathbf{X}(\mathbf{X}'\mathbf{G}^2\mathbf{X})^{-1}}{[n - \mathbf{g}'\mathbf{X}(\mathbf{X}'\mathbf{G}^2\mathbf{X})^{-1}\mathbf{X}'\mathbf{g}]} \right) \qquad (6.62)$$

where $\mathbf{g}$ is a vector whose elements are the diagonal elements of $\mathbf{G}$. The large-sample variance of $\hat{\kappa}$ is given by

$$Var(\hat{\kappa}) = 1/nA_1'(\hat{\kappa}) \qquad (6.63)$$

(see (3.48) for definition of A_1') and an estimated circular standard error of $\hat{\mu}$ by

$$\hat{\sigma}_{\hat{\mu}} = [(n-k)\hat{\kappa}A_1(\hat{\kappa})]^{-\frac{1}{2}} \tag{6.64}$$

from which a large-sample confidence interval for μ can be calculated (cf. (4.43)). With samples of fewer than 25 or 30 measurements, confidence intervals for the parameters are obtainable by bootstrap methods (§**8.4**) using parametric resampling.

In the case of a single predictor, (6.62) reduces to

$$Var\ \hat{\beta} = \frac{1}{\kappa A_1(\kappa) \sum\limits_{i=1}^{n} (v_i - \bar{v})^2} \tag{6.65}$$

where

$$v_i = g'(\beta x_i)x_i \tag{6.66}$$

The iterative procedure is easier to perform if the X have been centred at their means before fitting begins. One difficulty exists in the maximisation of this log-likelihood. When the true value of $\boldsymbol{\beta}$ is close to zero, the likelihood can have peaks not only near $\hat{\boldsymbol{\beta}} = \mathbf{0}$ but also as the elements of $\boldsymbol{\beta}$ become infinite, because use of any monotone link function also leads to a null model with mean direction close to $\mu \pm \pi$. In such a case (see Example 6.9) the estimate can arbitrarily be taken to be the location of the peak near zero.

With only one or two explanatory variables, a simpler estimation method is available. Set

$$\sum_{i=1}^{n} \cos(\theta_i - \mu - g(\boldsymbol{\beta}'\mathbf{x}_i)) = R\cos(\mu - \hat{\mu}) \tag{6.67}$$

Then the value of $\boldsymbol{\beta}$ which maximises R is also the value of $\boldsymbol{\beta}$ which maximises the log-likelihood. Having found the maximising $\boldsymbol{\beta}$, the values of $\hat{\mu}$ and $\hat{\kappa}$ can be found from (6.58)–(6.60). Inspecting the graph of R as a function of $\boldsymbol{\beta}$ will produce a good starting value for the numerical maximisation of R. Local maxima can exist quite close to the global maximum so that graphical exploration of the likelihood surface is advisable anyway.

To test hypotheses about $\boldsymbol{\beta}$, e.g. that $\boldsymbol{\beta} = \mathbf{0}$, the simplest approach is probably to determine whether the hypothesised value lies inside the confidence region determined for $\boldsymbol{\beta}$.

> **Example 6.9** For the data described in Example 6.3 and illustrated in Figure 6.3, we consider modelling the mean direction of movement of periwinkles as a function of distance moved. An alternative view of the

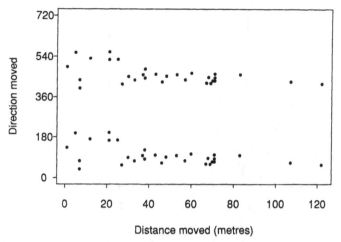

Fig. 6.10 Direction and distance moved by 31 small blue periwinkles: alternative display to Figure 6.3, with direction replicated on vertical scale. Comparison with Figure 6.9 suggests some evidence for dependence of direction on distance moved. See Example 6.9.

data is shown in Figure 6.10. Comparison with the prototype plots in Figure 6.9 suggests some dependence of direction on distance moved, and the possibility that dispersion may also depend on distance. For the moment, we ignore the latter possibility. To begin with, we can calculate the circular–linear correlation between direction and distance (cf. §6.2.1) as the extent of *circular–circular* association between $\Theta = direction$ and $\Phi = 2\tan^{-1}(distance)$ using, say, the correlation coefficient $\widehat{\rho}_T$ of §6.3.3. We get $\widehat{\rho}_T = -0.316$, which has a significance probability of 0.057 (based on a randomisation test using 1000 random permutations) when the hypothesis of no association between Θ and Φ is tested. So there is some evidence for association.

Next, we fit the model. A plot of the function R in (6.60) is shown in Figure 6.11, with a local maximum near 0 and, as noted above, possible maxima at $\beta = \pm\infty$. The iterative fit yields initial estimates $\widehat{\beta} = -0.0066, \widehat{\kappa} = 3.2$ and $\widehat{\mu} = 97.0°$. Having fitted the model, we can check the adequacy of fit with some diagnostic plots for the residual directions, after removing the estimated mean directions $\widehat{\mu} + \widehat{\beta}x_i$. Have we accounted for the dependence of the directions on distance? And do the residual directions appear to be a sample from a von Mises distribution?

Plots addressing these two questions are shown in Figures 6.12 and 6.13 respectively. Figure 6.12 suggests that the model has accounted for the relationship between the mean direction and distance moved, although the residual directions are clearly more dispersed for smaller distances. The von Mises Q–Q plot (cf. §4.5.3 (i)) is less sensitive to dispersion differences,

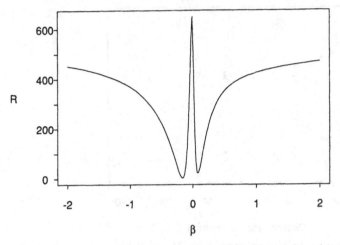

Fig. 6.11 Plot of R-component of Likelihood function as function of β, for data of Figure 6.10. See Example 6.9.

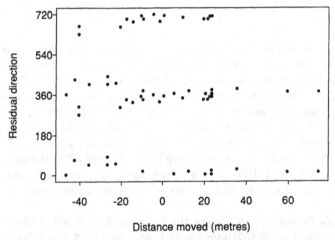

Fig. 6.12 Plot of residual directions from fitted mean direction regression model against direction moved, for data of Figure 6.10. There is some indication of dependence of dispersion on distance moved. See Example 6.9.

and there is little evidence of serious departure from a von Mises model. However, in view of the greater dispersion of the residuals at small distances, we shall postpone estimation of the precision of the parameter estimates until Examples 6.10 and 6.11, when the dispersion of direction will be modelled as well.

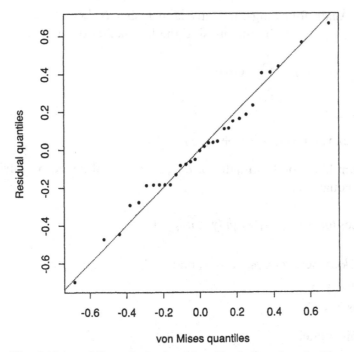

Fig. 6.13 von Mises Q–Q plot of residual directions in Figure 6.12. On the basis of this plot, the von Mises assumption for the errors does not seem unreasonable. See Example 6.9.

6.4.3 Circular regression for the concentration parameter

In the preceding discussion about modelling the mean direction, the von Mises assumption for the errors was not essential to fitting the model, at least in so far as estimation of the parameters μ and β was concerned. In what follows, such an assumption becomes essential, there being no natural dispersion parameter for circular data which suits the modelling in general.

We shall suppose that $(\theta_1, \mathbf{x}_1), \ldots, (\theta_n, \mathbf{x}_n)$ is a set of independent observations, with θ_i being drawn from a von Mises $VM(\mu, \kappa_i)$ distribution, $i = 1, \ldots, n$ (i.e. there is no relationship between the *mean direction* of Θ and the explanatory variables), and that κ_i is related to $\mathbf{x}_i$ by a link function of the form

$$h(\mathbf{x}) = h(\gamma' \mathbf{x}) \tag{6.68}$$

where

$$\gamma' \mathbf{x} = \alpha + \gamma_1 x_1 + \cdots + \gamma_k x_k \tag{6.69}$$

Here, we shall assume that the link function $h(x)$ is just the exponential function e^x. Again following the analysis in Fisher & Lee (1992), we fit the parameters $\mu, \alpha, \gamma_1, \ldots, \gamma_k$ by maximising the log-likelihood function

$$-\sum_{i=1}^{n} \log I_0(\kappa_i) + \sum_{i=1}^{n} \kappa_i \cos(\theta_i - \mu) \tag{6.70}$$

where

$$\kappa_i = \exp(\alpha + \gamma_1 x_{i1} + \cdots + \gamma_k x_{ik}) \tag{6.71}$$

The maximum likelihood estimates $\hat{\mu}, \hat{\alpha}, \hat{\gamma}_1, \ldots, \hat{\gamma}_k$ are obtained by solving the set of $k + 3$ equations

$$\sum_{i=1}^{n} [\cos(\theta_i - \mu) - A_1(\kappa_i)] h'(\gamma' \mathbf{x}_i) x_{ij} = 0 \tag{6.72}$$

$(j = 0, 1, \ldots, k)$, where $\alpha = \gamma_0, x_{i0} = 1$, and

$$R \sin \hat{\mu} = S \tag{6.73}$$
$$R \cos \hat{\mu} = C \tag{6.74}$$

where for this model

$$S = \sum_{i=1}^{n} \kappa_i \sin(\theta_i) \tag{6.75}$$

$$C = \sum_{i=1}^{n} \kappa_i \cos(\theta_i) \tag{6.76}$$

$$R = (S^2 + C^2)^{1/2} \tag{6.77}$$

For a given value of $\hat{\mu}$, the set of equations (6.72) can be solved iteratively. Given a starting value $\hat{\gamma}$, calculate an updated value $\hat{\gamma}^*$ from the updating equation

$$\mathbf{X}'\mathbf{W}\mathbf{X}(\hat{\gamma}^* - \hat{\gamma}) = \mathbf{X}'\mathbf{W}\mathbf{y} \tag{6.78}$$

where

$$y_i = [\cos(\theta_i - \mu) - A_1(\kappa_i)] / h'(\gamma' \mathbf{x}_i) A_1'(\kappa_i) \tag{6.79}$$

A_1' can be calculated using (3.48) and $\mathbf{W}$ is a diagonal matrix with elements

$$w_i = \{h'(\gamma' \mathbf{x}_i)\}^2 A_1'(\kappa_i) \tag{6.80}$$

So, the maximum likelihood estimates can calculated by picking a suitable starting value for $\hat{\gamma}$, computing $\hat{\mu}$ from (6.75) and (6.76), updating $\hat{\gamma}$ using (6.78), and repeating this cycle until convergence.

The large-sample variance of $\hat{\gamma}$ is given by

$$Var \; \hat{\gamma} = (\mathbf{X}'\mathbf{W}\mathbf{X})^{-1} \tag{6.81}$$

and an estimated circular standard error for $\hat{\mu}$ by

$$\hat{\sigma}_{\hat{\mu}} = \left(\sum_{i=1}^{n} \hat{\kappa}_i A_1(\hat{\kappa}_i) \right) -\frac{1}{2} \tag{6.82}$$

from which a large-sample confidence interval for μ can be calculated (cf. (4.43)). With samples of fewer than 25 or 30 measurements, confidence intervals for the parameters are obtainable by bootstrap methods (§**8.4**) using parametric re-sampling.

It is helpful, in practice, to have a graphical method of examining the validity of the (exponential) link function as a means of explaining possible dependence of dispersion on the covariates, and also to get starting values for the iterative solution of the maximum likelihood equations. The following simple procedure can be used with one covariate at a time. For a given covariate x, rearrange the data pairs (θ_i, x_i) so that the x_i are in increasing order, to get

$$(\theta_{(1)}, x_{(1)}), \ldots, (\theta_{(n)}, x_{(n)}) \tag{6.83}$$

say. For a small positive integer m (e.g. $m = 2$) calculate

$$\hat{\rho}_i = \text{resultant length of } \theta_{(i-m)}, \ldots, \theta_{(i+m)} \tag{6.84}$$

and then plot the $n - 2m$ points $(x_i, \log A_1'(\hat{\rho}_i))$. An approximately linear plot suggests that the exponential link function is providing a reasonable description of the relationship between the concentration parameter and the covariate. Starting values for $\hat{\alpha}$ and $\hat{\gamma}$ in the iterative estimation procedure can then be obtained by fitting a straight line to the plotted points and using, respectively, the estimated intercept and slope. If there are 'holes' in the sequence of x-values, care should be taken in selecting the corresponding set of θ for calculation of $\hat{\rho}$, because the method is based on the premise that, if the x-values are close together, then so are the corresponding κ-values.

Hypothesis tests concerning the parameters can be carried out as suggested in §**6.4.2**, by first calculating confidence regions for them and then checking whether these regions include the hypothesised parameter values.

Example 6.10 We shall continue the analysis begun in Example 6.9, of the periwinkle data of Example 6.3, by studying more closely the residuals from the fitted model for the mean directions. The link function diagnostic plot based on the exponential link is shown in Figure 6.14. On this basis, the

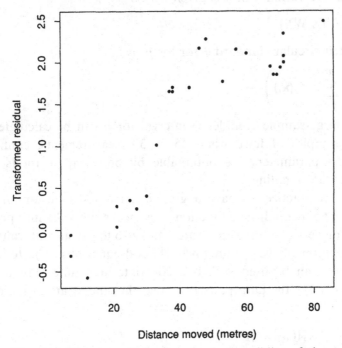

Fig. 6.14 Dispersion link plot, to check validity of the exponential function as a suitable dispersion link for regression of residual directions in Figure 6.12 on distance moved, for data of Figure 6.10. See Example 6.10.

exponential link appears adequate. The graph yields the starting values $\hat{\alpha}_0 = -0.36, \hat{\gamma}_0 = 0.04$ from which the final estimates $\hat{\alpha} = -0.0055, \hat{\gamma} = 0.034$ can be obtained using the iterative fitting method. We shall conclude the analysis of these data in Example 6.11.

6.4.4 Circular regression with a mixed model

In §6.4.2 and §6.4.3 we have studied ways of modelling either the mean direction or the concentration parameter as a function of some explanatory variables, separately. However, it is not uncommon to encounter data sets in which both mean direction and concentration seem to depend on these covariates. As can be seen in Examples 6.9 and 6.10, analysis of such data can be approached by two-stage fitting, i.e. by separate fitting of each type of model. However, the most efficient parameter estimates are obtained by simultaneous fitting of all parameters, if the full model is appropriate. We therefore present the equations for effecting this fit, for a combination of the two models previously considered. Note that a useful by-product of the two-stage fitting procedure is a good set of starting values for fitting the full model.

Set $\kappa_i = \exp(\gamma' x_i)$; then the log-likelihood function we seek to maximise is

$$-\sum_{i=1}^{n} \log I_0(\kappa_i) + \sum_{i=1}^{n} \kappa_i \cos(\theta_i - \mu - g(\beta' x_i)) \tag{6.85}$$

and it can be maximised by combining the methods of the previous two subsections.

Specifically, given starting values for $\widehat{\beta}$ and $\widehat{\gamma}$ (obtained, for example, by separate fitting as described above) we maximise (6.85) by first updating $\widehat{\beta}$ using a slightly modified version of (6.61), the matrix G^2 being replaced by GKG where K is a diagonal matrix with elements $\kappa_i A_1(\kappa_i)$. Also, in the definition of the vector y, $A_1(\kappa_i)$ replaces $A_1(\kappa)$. We then update the value of γ as in §**6.4.3**, except that (6.79) must be changed to

$$y_i = [\cos(\theta_i - \mu - g(\beta' x) - A_1(\kappa_i))]/(h'(\gamma' x_i)A_1'(\kappa_i)) \tag{6.86}$$

The weights (6.80) are unchanged. These two updating steps are alternated until convergence.

The estimates $\widehat{\beta}$ and $\widehat{\mu}$ are asymptotically uncorrelated with $\widehat{\gamma}$. The large-sample variance (6.81) is still correct, as is the estimated circular standard error for $\widehat{\mu}$ in (6.82); and a large-sample variance for $\widehat{\beta}$ can be estimated by

$$Var\,\widehat{\beta} = (X'GKGX)^{-1} \tag{6.87}$$

The comments in §**6.4.2** and §**6.4.3** concerning treatment of small samples and tests of hypotheses apply equally, here.

> **Example 6.11** We conclude the analysis of the periwinkle data of Example 6.3 which was pursued in Examples 6.9 and 6.10, by fitting the full mixed model. Using as starting values the estimates obtained from the two-stage process, we get: $\widehat{\mu} = 117.1°$, $\widehat{\sigma}_{\widehat{\mu}} = 0.0458$, $\widehat{\beta} = -0.0046$ with standard error (s.e.) 0.0013, $\widehat{\alpha} = 1.78$ (s.e. 0.25), $\widehat{\gamma} = 0.045$ (s.e. 0.009), and $\text{cov}(\widehat{\alpha}, \widehat{\gamma}) = -0.00007$. Since the magnitudes of $\widehat{\beta}$ and $\widehat{\gamma}$ are well in excess of their respective standard errors, both the mean component and the dispersion component of the regression model deserve inclusion; so the mixed model provides a better fit to the data than either of the two simpler models.
>
> *Note.* The values obtained for $\widehat{\beta}$ in this example and in Example 6.9 are half those given in Fisher & Lee (1992); in that paper, the link function $g(x) = 2\tan^{-1}(0.5x)$ was used, however the factor 0.5 is unnecessary.

References and footnotes. A number of regression models for an angular response, have been proposed in the literature. In particular, Gould (1969) suggested the 'barber's pole' model for the mean direction; see also Laycock (1975). Johnson & Wehrly (1978) suggested some simple models for the mean direction or dispersion which were subsequently extended by Fisher & Lee (1992).

6.4.5 *Circular regression with a circular covariate*

Little work has been done in this area to date. In an unpublished report in 1986 entitled 'Projective angular regression', T.D. Downs gave a method of fitting a model for the mean direction of a von Mises random variable Θ given a value ϕ of a von Mises covariate Φ, using a link function method similar to that described above. Subsequently, Fisher & Lee (1992) suggested that when the distributions of Θ and Φ are not too dispersed, (e.g. when each has a value of the von Mises concentration parameter of at least 2), then the regression problem could be handled satisfactorily by transforming the data to linear variates, as is done in Chapter 7. See Laycock (1975) for an alternative approach to the problem.

7

Analysis of Data with Temporal or Spatial Structure

7.1 Introduction

We shall be concerned with three types of data in this chapter:

(I) Circular observations in relation to a time sequence $t_1, t_2, \ldots$
(II) Circular observations in relation to a one-dimensional spatial sequence $x_1, x_2, \ldots$, in other words a sequence of points along a line.
(III) Circular observations in relation to a corresponding set of two-dimensional spatial locations, in other words positions in the plane.

Time-dependent sequences of data of type I arise commonly in Meteorology and Oceanography as measurements of wind or current direction at a particular location; often, but not always, each direction has an associated measurement of strength, so that the data constitute velocities. When the measurements have been recorded at equal intervals (e.g. hourly or daily measurements of wind direction), the sequence is called a *time series*. Typically, the questions of interest relate to finding models which capture a sufficient amount of the dependence structure to allow future behaviour of the series to be predicted.

Data sequences gathered along a line (type II) can occur in Structural Geology, for example, when mapping along a mining tunnel. Here, interest centres on extracting an overall trend, or detecting sudden changes in the general trend of the sequence. Data of this type differ from those of type I in that they can be looked at in reverse order – the order depended simply on the direction $x_1, x_2, \ldots$ or $x_n, x_{n-1}, \ldots$ used for sampling – whereas time-ordered sequences usually have only one sensible direction. Notwithstanding this distinction, data sets of type II are commonly referred to as time sequences or series, and we shall do so here.

The problem of analysing circular data with a spatial aspect (type III) is ubiquitous in Structural Geology. In Chapter 1, an illustration was given of

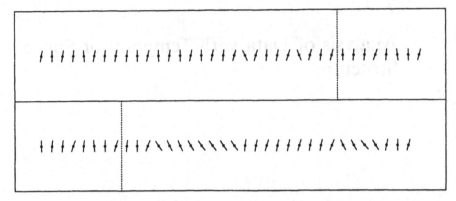

Fig. 7.1 Needle-plot of median trends of samples of five cleat trend measurements taken at 20-metre intervals along a tunnel in Wallsend Borehole Colliery, NSW. Dashed lines indicate over lapping sections of the series. It is of interest to detect changes in mean trend, which appear to be good indicators of potentially hazardous mining conditions ahead. See Example 7.1.

the difficulties of sampling and interpreting multimodal data from a single sampling site (see Figure 1.4); yet this is but the atomic part of the general problem of distilling the information about structural domains from data sampled throughout a region – the number of differently-oriented fracture sets present in the region as a whole, the parts of the region in which they occur, which are the dominant sets, and so on. Other applications occur in assessing stress fields, as described by Hansen & Mount (1990):

> Upper crustal (< 20 km) tectonic stress directions are inferred from borehole elongations (breakouts), earthquake focal mechanisms, hydraulic fracture orientations, alignments of young volcanic flank eruptions, and recent slip on faults. Shallow stress measurements (0–5 km) inferred from borehole and geologic measurements tend to be consistent with the deeper stress directions (5–20 km) inferred from earthquake focal mechanism studies ... The consistency between shallow and deep stress orientation measurements indicates that an upper crustal stress field is being measured.
>
> Stress orientation measurements are commonly plotted as short bars on a map of a region under study, with the bars indicating the direction of maximum, or minimum, stress at the point of measurement ... Available stress data are subject to errors arising from a variety of sources, and within a region there may exist large areas containing few, or no, measurements.

Spatial smoothing and interpolation of the stress measurements can be used to infer stress fields and stress trajectories (a stress trajectory being a curve whose tangent at any point is in the direction of one of the principal stresses at that point).

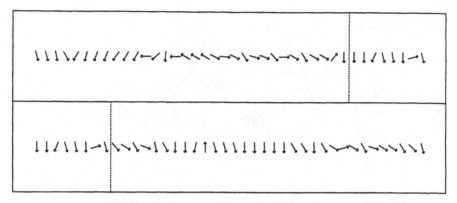

Fig. 7.2 Hourly measurements of wind direction, for three days, collected at Black Mountain, ACT Australia. It is of interest to see whether a satisfactory model can be fitted, which would then allow reliable predictions to be made of future wind directions. See Example 7.2.

Example 7.1 Figure 7.1 shows a 'needle-plot' of a time-sequence of measurements of face-cleat from the Wallsend Borehole Colliery, NSW. The data were obtained by sampling at 20-metre intervals along a traverse of a coal seam, taking a sample of five measurements from each sampling site, and selecting the median axis of each sample. In many coal mines, a shift in the preferred direction of the dominant set is a very good predictor of hazardous mining conditions due to forthcoming faults or dykes. A simple method is needed to highlight such changes, so that data can actually be gathered and analysed underground to give timely notice of hazards. (The data are listed in Appendix B22.)

Example 7.2 Figure 7.2 shows a time series of 72 wind directions, comprising hourly measurements for three days at a site on Black Mountain, ACT, Australia. The data were collected as part of an experiment to calibrate three anemometers, as reported in Cameron (1983). (They are listed in Appendix B23.) It is of interest to know whether the data can be modelled adequately by a reasonably simple model, so that future wind patterns can be forecast.

Example 7.3 Figure 7.3 shows the result of a lineament interpretation from a remotely sensed image, collected as part of a project to identify and map fracture sets in the Palm Valley, NT region which may control the flow of natural gas. Each line corresponds to a linear feature in the image which the interpreter believes to be of geological significance. Thus a short line may correspond to a small ridge running down to a river, a long line to the ridge-line of a complete chain of hills. One question of interest is to identify the different fracture sets present, and to highlight where they occur.

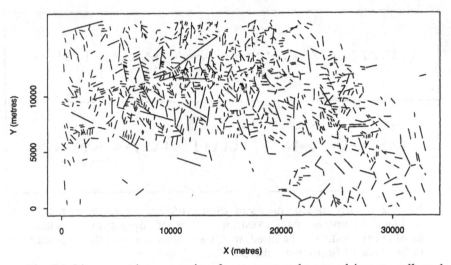

Fig. 7.3 Lineament interpretation from a remotely sensed image, collected as part of a project to identify and map fracture sets in the Palm Valley, NT region which may control the flow of natural gas. One question of interest is to identify the different fracture sets present, and to highlight where they occur. See Example 7.3.

The lineament interpretation was performed by Mr M. Chinkul (University of New South Wales), whose permission to use the unpublished data is gratefully acknowledged.

In Section 7.2, we consider methods for analysing circular data with a time-dependent character, and in Section 7.3 some suggestions are made for exploring data with spatial structure.

7.2 Analysis of temporal data

7.2.1 Introduction

As in other parts of this book, we begin with some simple exploratory methods for revealing structure in the data. These are described in §7.2.2. Tests for changes in trend, and for serial dependence of a sequence of circular measurements, are given in §7.2.3. Finally, the problem of fitting formal models to time series, analogously to that for linear data, is dealt with in §7.2.4.

7.2.2 *Exploratory analysis*

Let $\theta_1, \ldots, \theta_n$ be a sequence of circular measurements (vectors or axes) made at time points $t_1, \ldots, t_n$, where $t_1 < \cdots < t_n$. Typically, but not always, the time points will be equally spaced (e.g. hourly temperature, wind speed daily at 6.00am, equally-spaced measurements along an excavation tunnel). A basic concept of exploratory analysis is to extract and display the broad structure in the data, and this can be done largely with readily available algorithms. We consider separately methods for elucidating general trends in the data, and for highlighting possible dependence of each point θ_i in a circular time series on preceding values.

(i) Highlighting trends

Set $x_i = \cos \theta_i$, $y_i = \sin \theta_i$, $i = 1, \ldots, n$. The basic approach we shall adopt to smoothing the data sequence is to smooth the x- and y-sequences *separately but identically*, and then to recombine them into a smoothed sequence of circular measurements (personal communication from G.S. Watson). There are numerous methods for smoothing a sequence of linear measurements $x_1, x_2, \ldots$ which, although specifically designed for independent data, are useful for exploring correlated sequences as well. For a general discussion see Härdle (1990).

One simple class of methods comprises the so-called *kernel* methods, akin to the nonparametric density estimation technique described in §**2.2**(*iv*). The kernel estimate of $x(t)$ [$y(t)$] at a point t is simply a weighted average of the values of x_i [y_i] measured at times t_i near t. Specifically,

$$\widehat{x}(t) = \frac{\sum_1^n K_h(t - t_i) x_i}{\sum_1^n K_h(t - t_i)} \tag{7.1}$$

and

$$\widehat{y}(t) = \frac{\sum_1^n K_h(t - t_i) y_i}{\sum_1^n K_h(t - t_i)} \tag{7.2}$$

where

$$K_h(t) = h^{-1} K(t/h) \tag{7.3}$$

The smoothed estimate of $\theta(t)$ is then

$$\widehat{\theta}(t) = \tan^{-1}\{\widehat{y}(t)/\widehat{x}(t)\} \tag{7.4}$$

Typically, h will depend on the sample size n (cf. the remarks in §**2.2**(*iv*)); see Härdle (1990, Chapter 3) for more discussion about this point and computational aspects. Here, we shall be content to recommend that a range

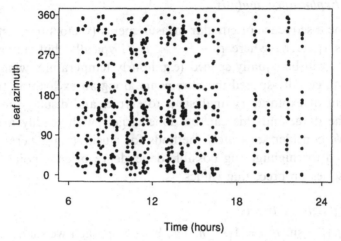

Fig. 7.4 Azimuths of leaf tips in response to direction of solar beam, with the mean direction centred at 180°. No replication of the angles is necessary because the azimuths are confined to a subset of the circle. See Example 7.4.

of values of h be used, to study both small-scale and large-scale trends in the data. As with density estimation, the choice of the kernel function K is, broadly speaking, less important than the choice of h in governing the result of the smoothing operation. A common choice for the function $K(t)$ is the Gaussian kernel

$$K(t) = (2\pi)^{-\frac{1}{2}} \exp(-t^2/2) \tag{7.5}$$

Whilst simple to implement, kernel-based smoothers have a tendency to clip (i.e. underestimate) peaks in the data, and correspondingly to fill in troughs. The implication of this in the present context is that different directions may be smoothed differently. One way of avoiding this problem is to use a smoothing method which does not have such side-effects. The local quadratic version of the so-called 'loess' algorithm (see Cleveland & Grosse, 1991) is suitable in this respect. It effectively performs a local regression in the neighbourhood of a point at which a smoothed value is sought. The number of neighbouring points is controlled by the *span* of the regression, expressed as a percentage of the size of the data set. Details of the method and of how to get access to computer code are given in this reference. We shall illustrate its use in Examples 7.4 and 7.6 with temporal data and then in Example 7.9, with spatial data.

Example 7.4 The data in Figure 7.4 come from a series of experiments studying the orientations of leaves from a range of plants to the solar beam (Herbert 1983, Herbert & Larsen 1985). The leaves under study (from the plant species *Calathea*) were generally unshaded for a 4–5-hour period around noon, which is the period of particular interest. (In a personal communication, Dr Herbert observed: 'Unfortunately, lunch was served at the field station in Panama every day at exactly 12 noon. If one was late, then there might be no food left. So, there are fewer measurements right near noon.') The plotted measurements were gathered from a number of leaves over an 8-day observation period and combined *modulo* 24 hours to provide an overall picture of the daily pattern. Thus, the data set is an amalgam of several interleaved time-sequences so that formal statistical inference would not be easy. Nevertheless, it makes sense to use exploratory analysis to highlight any trends which may be present. The measurements are the azimuth (compass direction) in which the leaf tip points. Because of the size of the data set, the data have been displayed in the kind of rectangular plot used in §6, with the difference that after they have been centred at 180°, the duplication of angles has not been necessary.

Two different smoothed versions of the data are shown in Figure 7.5, using different spans (50% and 30%) for the local regression. If we focus on the period of particular interest around noon, the latter smooth reveals some interesting structure between 10am and 3pm. Some important information appears to have been lost as a consequence of the lunch break.

Choosing the correct amount of smoothing can be aided using plots of residuals from the fitted curve, and then smoothing these to detect residual structure. We shall not pursue this here.

References and footnotes. An alternative approach, using an analogue of splines for linear data, has been suggested by Watson (1985).

(ii) Dependence structure of a series

If the data $\{\theta_i\}$ constitute a time series, it is usually of interest to examine the dependence structure after removing overall trends of the sort described in the discussion above. Again, we shall look at direct analogues of methods for linear time series, using the circular correlation coefficient ρ_T described in §6.3.3. For any integer $k = 1, 2, \ldots$, we define the sample lag–k (circular) correlation coefficient of $\{\theta_i\}$ to be the value, $\hat{\rho}_{T,k}$ say, of $\hat{\rho}_T$ calculated from the data pairs

$$(\theta_1, \theta_{k+1}), \ldots, (\theta_{n-k}, \theta_n)$$

A circular correlogram can then be obtained by plotting the points $(k, \hat{\rho}_{T,k})$, $k = 1, 2, \ldots$. This correlogram can be interpreted similarly to a linear correlogram. An illustration of the use of this technique is given in Example 7.8, in §7.2.4.

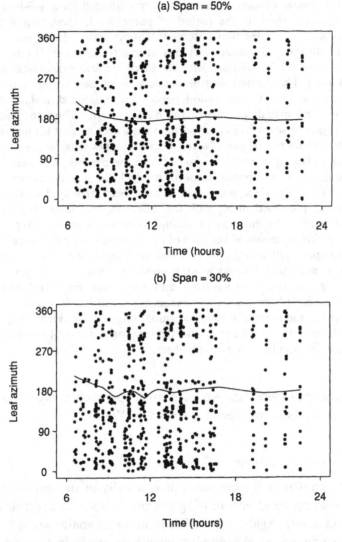

Fig. 7.5 Two smoothed versions of the leaf azimuth data in Figure 7.4. (a) local regression with 50% span. (b) local regression with 30% span. See Example 7.4.

7.2.3 Detecting changes of direction and serial dependence

(i) Detecting change of preferred trend

We begin with a simple graphical approach which has found useful application in coal mining. Suppose that the measurements have been made at equally-spaced time intervals, or at equally-spaced intervals along a linear traverse. The statistical model we have in mind is that the measurements $\{\theta_i\}$

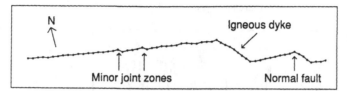

Fig. 7.6 SLIME plot for the cleat orientations plotted in Example 7.1. Changes in direction of the plot are useful forewarnings of hazardous mining conditions. See Example 7.5.

are *consistent within blocks*, that is, the series consists of a number of blocks of data, with the data in each block or zone being a sample of independent measurements from a common population, but with the population mean directions or orientations differing from block to block. A cumulative sum (or CUSUM) plot can be used to explore this possibility. Let

$$C_j = \sum_1^j x_i, \quad S_j = \sum_1^j y_i, \quad j = 1, 2, \dots \tag{7.6}$$

and plot the points $(C_1, S_1), (C_2, S_2), \dots$. Changes of block should show up in the resulting graph as marked changes of direction in the CUSUM plot.

Example 7.5 We apply this method to the sequence of measurements of face-cleat in Example 7.1. The median direction is used rather than the mean direction because the method is designed for use *in situ*; the median direction can be found without the use of a calculator (often not allowed in underground coal mines) and the plot updated immediately, allowing a geologist to monitor cleat patterns in the mine in a timely fashion. The result is shown in Figure 7.6, augmented with some information about structural features which were subsequently discovered. As can be seen from the plot, changes of direction in the so-called SLIME plot (for Sequential LInked MEdian) are helpful indicators of possibly hazardous mining conditions in several instances, allowing special precautions to be taken when mining in the new zone following a direction change.

Note that the SLIME plot is based on the values of the data themselves, and preserves the local directions or trends at each point on the graph. Whilst the actual local direction is highly relevant in this context, it may not be in other applications. There, the fact that the SLIME plot can wander off the page or screen as more data accrue may be a nuisance. Such is not the case with the method to be presented next, which is designed to be used retrospectively (after all the data are to hand) rather than as the data are collected. It leads to a formal test which can be helpful in situations which

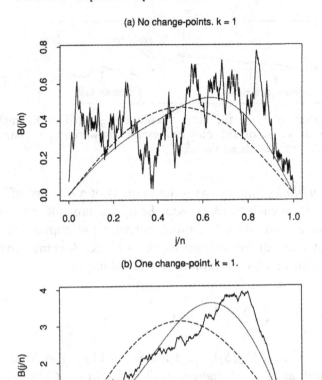

Fig. 7.7 Prototype rank CUSUM plots, with smoothed CUSUMS based on component k (dashed line) and components k and $k+1$ (dotted line).

are less clear-cut than the cleat-monitoring data of Example 7.5. The method is based on using the ranks of the data rather than the data themselves.

Define $\theta_0 = 0$, and determine the circular ranks $\gamma_0, \gamma_1, \ldots, \gamma_n$ (cf. (2.35)) corresponding to the $n+1$ values $\theta_0, \theta_1, \ldots, \theta_n$. For $j = 1, \ldots, n$ calculate

$$U_0 = 0, \quad U_j = (2/n)^{\frac{1}{2}} \sum_{i=1}^{j} \cos \gamma_i \tag{7.7}$$

$$V_0 = 0, \quad V_j = (2/n)^{\frac{1}{2}} \sum_{i=1}^{j} \sin \gamma_i \tag{7.8}$$

(c) Two change-points. k = 2

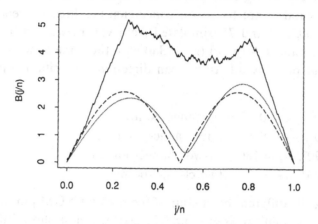

(d) Two change-points: second type. k = 1

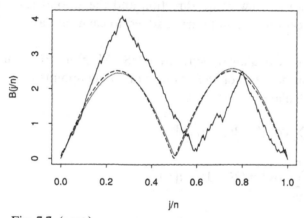

Fig. 7.7 (*cont.*)

and then

$$B(0) = 0, \quad B\left(\frac{j}{n}\right) = (U_j^2 + V_j^2)^{\frac{1}{2}}, \quad j = 1,\ldots,n \qquad (7.9)$$

A rank CUSUM plot can be obtained by plotting the points

$$\left(\frac{j}{n}, B\left(\frac{j}{n}\right)\right), \quad j = 0,\ldots,n$$

as a line plot. A sustained change of direction of the graph is interpreted as an indication of a change in direction of the mean direction or trend of the series. To get an idea of how this plot behaves in response to different forms of data sequences, some data have been simulated from four different models and the corresponding rank CUSUM graphs calculated, as shown in Figure 7.7. The models are based on concatenating three sequences of respective lengths 100, 200 and 75, simulated from von Mises distributions $VM(\mu_1, \kappa)$, $VM(\mu_2, \kappa)$ and $VM(\mu_3, \kappa)$ (cf. §3.3.6) with the same concentration parameter $\kappa = 2$ but possibly differing mean directions. Specific parameter choices were:

(a) $\mu_1 = 0°$, $\mu_2 = 0°$, $\mu_3 = 0°$ (no change-point)
(b) $\mu_1 = 0°$, $\mu_2 = 0°$, $\mu_3 = 90°$ (one change-point)
(c) $\mu_1 = 0°$, $\mu_2 = 90°$, $\mu_3 = 180°$ (two change-points)
(d) $\mu_1 = 0°$, $\mu_2 = 90°$, $\mu_3 = 0°$ (two change-points)

The characteristically different behaviour of the rank CUSUM plot for the different generating mechanisms is evident. The last two plots show differing behaviours for sequences of different types, the first changing from one new mean direction to another new mean direction and the second reverting to the original mean direction. Smooth curves added to each plot are explained below.

The next step is to calculate a smooth CUSUM plot, to be used in conjunction with a formal test for the number of change-points.

For $k = 1, 2, \ldots$, calculate

$$X_k = (2/n)^{\frac{1}{2}} \sum_{i=1}^{n} \cos [k\pi(2i - 1)/2n] \cos \gamma_i \qquad (7.10)$$

$$Y_k = (2/n)^{\frac{1}{2}} \sum_{i=1}^{n} \cos [k\pi(2i - 1)/2n] \sin \gamma_i \qquad (7.11)$$

and then

$$Z_k = (X_k^2 + Y_k^2)^{\frac{1}{2}}, \quad k = 1, 2, \ldots \qquad (7.12)$$

The sizes of the components $Z_1, Z_2, \ldots$ are indicative of the number of change-points in the series.

Critical values. Let n_0 be the maximum number of change-points deemed possible, given the length of the series n, and denote by Z_{max} the maximum of $Z_1, \ldots, Z_{n_0}$. Then, for large n, we have, approximately,

Prob $(Z_{max} > \gamma$ assuming no change-points)
$$= 1 - [1 - \exp(-\gamma^2/2)]^{n_0} \qquad (7.13)$$

from which a suitable value for γ (depending on n_0) can be found to make the right hand side of (7.13) equal to some desired value (e.g. equal to 0.05).

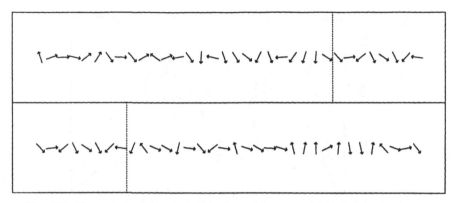

Fig. 7.8 Time series of azimuths of position at which a flare commences burning. The dotted lines show portions of overlap for the two panels. It is of interest to detect changes in mean direction in the series. See Example 7.6.

Suppose that the test identifies m of the n_0 components, $Z_{k_1},\ldots,Z_{k_m}$ say, as being significant (in the sense that each exceeds the value of γ determined from (7.13)). As a further aid to deciding on the number of change-points, a smooth rank CUSUM plot can be created based on these significant components. For $i = 1,\ldots,m$, calculate

$$U_{k_i}^* = 2^{\frac{1}{2}} U_{k_i}/(k_i \pi) \tag{7.14}$$

$$V_{k_i}^* = 2^{\frac{1}{2}} V_{k_i}/(k_i \pi) \tag{7.15}$$

$$B_U(y) = \sum_{i=1}^{m} U_{k_i}^* \sin(k_i \pi y) \tag{7.16}$$

$$B_V(y) = \sum_{i=1}^{m} V_{k_i}^* \sin(k_i \pi y) \tag{7.17}$$

and then the function

$$B^*(y) = (B_U^2(y) + B_V^2(y))^{\frac{1}{2}} \tag{7.18}$$

Finally, plot the points

$$\left(\frac{j}{n}, B^*\left(\frac{j}{n}\right)\right), \quad j = 0,\ldots,n$$

as a line plot. The smooth lines added to the plots in Figure 7.7 are just the smoothed rank CUSUMs based on the first, and the first and second, components.

Example 7.6 Figure 7.8 shows a time series of measurements obtained from an experiment to assess the relative stability of flare–projectile assemblies.

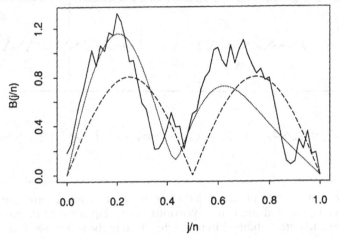

Fig. 7.9 Rank CUSUM plot for the flare data in Figure 7.8, with smooth CUSUMs based on component 2 (dashed line), and components 2 and 3 (dotted line), added. Comparison with Figure 7.7(d) suggests two change-points. See Example 7.6.

Table 7.1. *The first six components of the smoothed rank CUSUM, for detecting the number of change-points in the flare–azimuth data sequence.*

Statistic	\multicolumn{6}{c}{Component number k}

Statistic	1	2	3	4	5	6
X_k	−0.76	3.11	2.84	0.16	1.15	−0.49
Y_k	−0.12	1.76	0.40	0.08	0.53	−1.04
Z_k	0.77	3.58	2.87	0.18	1.27	1.15

(The data are listed in Appendix B24.) A flare, attached to a projectile, is launched upward from a launch point O in a fixed direction. At some point P in space, the flare commences burning. The azimuth of P relative to O gives an indication of the variability of the assembly as more and more trials are conducted with it. The data shown are based on 60 successive launches (see Lombard (1986) for a more detailed discussion).

The rank CUSUM plot is shown in Figure 7.9. Comparison with the prototype plots in Figure 7.7, particularly Figure 7.7(d), already suggests the distinct possibility of at least two change-points. The values of the components Z_k, $k = 1, \ldots, 6$ are given in Table 7.1. If we take the (conservative) view that we shouldn't expect to detect more than $m = 6$ change-points in a sequence of length 60, we find, from (7.13), that the value $\gamma = 2.86$ corresponds to the 10% critical point, and $\gamma = 3.54$ to the 1% point. On

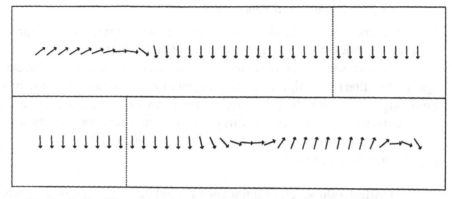

Fig. 7.10 Heavily smoothed version of flare data in Figure 7.8, using *loess* and a span of 0.5. (Again, the dotted lines show portions of overlap for the two panels.) Two change-points are evident, with the suggestion of a third at the end of the series. See Example 7.6.

this basis, the second component Z_2 is clearly significantly large, and the third component perhaps marginal. To clarify the issue, we construct the smooth rank CUSUMs based on just Z_2, and also Z_2 and Z_3 together. These are shown in Figure 7.10. It is clear that the general shape of the raw rank CUSUM is being captured by the second component alone, and that the addition of the third component serves only to improve the 'two-bump' approximation, without adding qualitatively different feature. So we conclude that there are two change-points in the data sequence.

We note two additional points arising from this example.

Firstly, the objective of the smoothed rank CUSUMs is *not* to get a good approximation to the rank CUSUM (which could be done simply by taking a large number of components), but rather to capture the qualitative features of the rank CUSUM with a minimal number of components.

Secondly, it is interesting to see whether a smoothing method can help in eliciting information about change-points. We use the *loess* method mentioned earlier, with a span (effectively, a bandwidth) which encompasses 50% of the data. The result is shown in Figure 7.10, and is very striking: there is strong evidence for the existence of two change-points and, just possibly, for the existence of a third at the end of the sequence, although the data are too few here to be sure. The markers at the smoothed values correspond to the 12th and 42nd points of the sequence, which were identified by Lombard (1986) as the change-points in his original analysis of these data.

References and footnotes. The SLIME plot technique was developed by Shepherd & Fisher (1981, 1982). The rank CUSUM methods are due to Lombard (1986, 1988).

(ii) Detecting serial dependence

It is sometimes of interest to establish whether a time series of measurements is in fact a series of independent and identically distributed measurements (i.e. a random sample of measurements in which time order is irrelevant) against the alternative that each measurement is correlated with one or more preceding measurements. This alternative form of association is known as *serial* correlation. It is easy to carry out a formal test for association with the immediately preceding measurement, by using the first lag correlation $\hat{\rho}_{T,1}$ defined by equation (7.6).

Critical values. The significance probability of $\hat{\rho}_{T,1}$ can be evaluated using a permutation test (cf. §8.5). For very short series, it is feasible to enumerate all $n!$ permutations of the data, and to compute $\hat{\rho}_{T,1}$ for each. For $n \geq 10$, say, the permutation distribution can be approximated by calculating $\hat{\rho}_{T,1}$ for a large number of random permutations of the series.

> **Example 7.7** For the wind direction data of Example 7.2, we get the sample lag 1 correlation to be 0.49. On the basis of just 100 random permutations of the data, it is clear that this is a significantly large value of $\hat{\rho}_{T,1}$, as it exceeds in magnitude all other 100 correlations in the approximate permutation distribution. So the null hypothesis that the measurements are independent seems very implausible.

References and Footnotes. The first test for serial correlation was introduced by Watson & Beran (1967), using an unsigned measure of association; computational aspects were developed by Epp *et al.* (1971), including a large-sample Normal approximation. No such approximation is currently available for the procedure given above; however, the randomisation test should be satisfactory in most situations.

7.2.4 Fitting autoregressive models to time series

Having carried out initial exploration of a time series, it is sometimes of interest to seek a parametric model which fits the series. If such a model can be found, we can then predict future values of the series.

Circular time series are rather more difficult to handle than linear time series, because of the lack of convenient probability models for correlated circular data. Also, they are more difficult to explore by graphical means, when the measurements are not reasonably concentrated around the mean trend of the series. For example, detecting a cyclic component in a dispersed circular time series is a difficult graphical exercise.

Our approach to model fitting will therefore take one of two forms, depending on whether the series appears to be 'noisy' or not – that is, depending on how much variation there appears to be around the overall trend of the series. For noisy series, we shall attempt to fit models to the circular data themselves, whereas for less noisy series we shall transform the data to linear data, and take advantage of the wealth of methods and software available for linear time series.

The general approach in either case is similar to that employed for angular regression (§**6.4**) in that it utilises link functions. Suppose that $g(x)$ is an increasing function of x which transforms values from $(-\infty, \infty)$ to the circle $(-\pi, \pi)$, and such that $g(0) = 0$ and $g(-x) = -g(x)$ for all values of x. For example, we could use a link function similar to that used in §**6.4**, namely

$$g_T(x) = 2\tan^{-1}(x) \tag{7.19}$$

Also, define the function

$$G(\theta; \omega) = g(\omega g^{-1}(\theta)) \tag{7.20}$$

Then G is a function which transforms values from the circle back to the circle.

For such link functions, if X is a linear random variable, then

$$\Theta = g(X) \tag{7.21}$$

is a circular random variable, and conversely, if Θ is a circular random variable,

$$X = g^{-1}(\Theta) \tag{7.22}$$

is a linear random variable.

Define θ_i to be a linked $ARMA(p, q)$ process (i.e. a *LARMA* process) if its *linked linear process* $g^{-1}(\theta_t)$ is an $ARMA(p, q)$ process. We shall not give detailed explanations of $ARMA(p, q)$ processes here, beyond some heuristic definitions; the reader is referred to a standard time series text (e.g. Chatfield 1989) for details about Box–Jenkins models. Briefly, an autoregressive linear series of order p (or $AR(p)$) $X_1, X_2, \ldots$ is one in which

$$X_t = \alpha_1 X_{t-1} + \cdots + \alpha_p X_{t-p} + \epsilon_t, \quad t = p+1, \ldots \tag{7.23}$$

where $\epsilon_1, \epsilon_2, \ldots$ is a sequence of independent errors. The analogy with linear regression is clear. Again, a moving average linear series of order q (or $MA(q)$) is one in which

$$X_t = \beta_1 \epsilon_{t-1} + \cdots + \beta_q \epsilon_{t-q} + \epsilon_t, \quad t = q+1, \ldots \tag{7.24}$$

An $ARMA(p,q)$ process is simply a combination of these two processes. We shall concentrate largely on AR processes. In fact, we shall look at two ways ($LAR(p)$ processes and $CAR(p)$ processes) for modelling autoregressive behaviour in a circular time series, depending on whether the series is dispersed or concentrated.

A circular $AR(p)$ process (or $CAR(p)$ process†) can be defined directly, as one for which the distribution of Θ_t, given the past values $\Theta_{t-1} = \theta_{t-1}, \Theta_{t-2} = \theta_{t-2}, \ldots$, is von Mises $VM(\mu_t, \kappa)$ (although any other unimodal circular distribution symmetric about its mean could be used) where the mean direction μ_t is given by

$$\mu_t = \mu + g\left[\alpha_1 g^{-1}(\theta_{t-1} - \mu) + \cdots + \alpha_p g^{-1}(\theta_{t-p} - \mu)\right] \tag{7.25}$$

and κ is a constant concentration parameter. Once the parameters in (7.25) have been estimated, the mean of the next value in the series can be forecast.

Fitting CAR(p) models for dispersed series ($\kappa < 2$) This case is tractable for small values of p. The parameters are estimated by maximising

$$\prod_{t=p+1}^{n} f\left\{\theta_t - \mu - g\left\{\alpha_1 g^{-1}[(\theta_{t-1} - \mu)/2] \cdots + \alpha_p g^{-1}[(\theta_{t-p} - \mu)/2]\right\}\right\}$$

$$\times \prod_{t=1}^{p} f(\theta_t - \mu) \tag{7.26}$$

where f denotes the probability density function of the von Mises $VM(0, \kappa)$ (see (3.31)). This minimisation is straightforward, using the Newton–Raphson algorithm. Full details of its implementation are given in Fisher & Lee (1993).

Fitting LARMA(p,q) models for concentrated series ($\kappa \geq 2$) For any value of μ, transform the data to linear data using the chosen link function. A suitable one for such data is the so-called *probit* link, whose inverse is given by

$$x(\mu) = \Phi^{-1}((\theta - \mu)/2\pi + \tfrac{1}{2}) \tag{7.27}$$

where the value of $\theta - \mu$ is taken to lie between $-\pi$ and π. This can be used to obtain values $Y_1(\mu), \ldots, Y_n(\mu)$ say. Linear methods can then be applied to maximise the likelihood of $Y_1(\mu), \ldots, Y_n(\mu)$, which is then optimised as a function of μ. A suitable starting value for μ is the sample mean direction of $\theta_1, \ldots, \theta_n$.

† Called 'inverse autoregressive processes', or $IAR(p)$, by Fisher & Lee (1993)

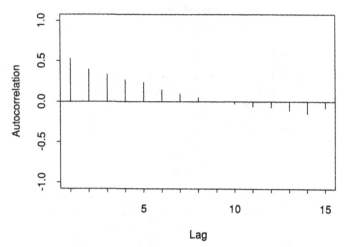

Fig. 7.11 Circular autocorrelations for the time series of wind directions in Figure 7.2. The manner in which the correlations fall away suggests that a $CAR(1)$ process may be an appropriate model. See Example 7.8.

So we have two ways of modelling autoregressive behaviour in a circular time series, depending on the degree of 'noise' in the series. Note that because they are different models their parameters have differing interpretations, and so cannot be expected to have comparable values.

Example 7.8 We shall illustrate the two methods by applying them to the wind direction data in Example 7.2. The analysis follows that described in Fisher & Lee (1993).

We begin by looking at the circular autocorrelation function. This is shown in Figure 7.11. Based on prototype autocorrelation plots in Fisher & Lee (1993), there is evidence that a $CAR(1)$ model may be appropriate. For model fitting, starting values for μ and κ can be obtained by treating the time series as a random sample and calculating point estimates as in §**4.5.5**, and a starting value, for $\alpha_1 = \alpha$ say, can be found from the prototype plots. We get

$$\hat{\mu}_0 = 291.2°, \quad \hat{\kappa}_0 = 1.9, \quad \hat{\alpha}_0 = 0.5.$$

This leads to the parameter estimates

$$\hat{\mu} = 289.5°, \quad \hat{\kappa} = 2.5, \quad \hat{\alpha} = 0.68$$

The circular standard error of $\hat{\mu}$ is then 0.086 and standard errors for $\hat{\kappa}$ and $\hat{\alpha}$ are 0.352 and 0.138 respectively. Estimated mean directions of future values in the series can then be obtained from (7.25).

In view of the fact that $\hat{\kappa}$ exceeds 2.0, we consider using the linked method as well. Plots of the transformed series using the probit link, and of the autocorrelation and partial autocorrelation functions are shown in

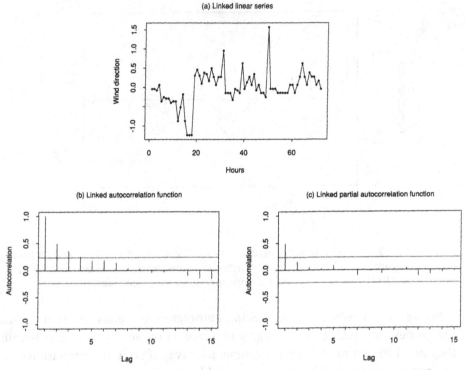

Fig. 7.12 Analysis of the wind direction time series in Figure 7.2 using probit link to fit an *LAR*(1) model. The autocorrelation and partial autocorrelation functions suggest that an *LAR*(1) model may be reasonable. (a) Linked linear time series. (b) Autocorrelation function for linked series. (c) Partial autocorrelation function for linked series. The dotted lines in (b) and (c) are an indication as to which correlations are significantly different from zero. See Example 7.8.

Figure 7.12. The dotted lines at $\pm 2n^{-\frac{1}{2}}$ provide an approximate test for non-zero correlations, at the 5% level. The two correlation plots taken together suggest that an *LAR*(1) model may be suitable (see Chatfield, 1989, for use of these plots to infer the order of the autoregression). The parameter estimates for this model are then

$$\hat{\mu}_0 = 297.2°, \ \hat{\alpha} = 0.52, \ \hat{\sigma}^2 = 0.146$$

The standard error of $\hat{\alpha}$ is 0.10. (A circular standard error for $\hat{\mu}$ could be found by transforming the residuals from the fitted model back to the circle and calculating the usual circular standard error from equation (4.21) with n reduced by the total number of fitted autoregressive and moving-average parameters, here 1.) Forecasting under this model proceeds using standard linear time series methods.

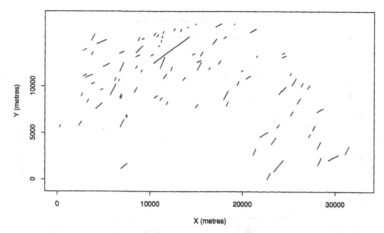

Fig. 7.13 The subset of lineaments in Figure 7.3 which lie in a 45°-arc around the *NE–SW* axis. See Example 7.9.

References and footnotes. The two methods are described in more detail in Fisher & Lee (1993), where a comparison with other approaches is also given. One of the other approaches is an extension of a method proposed by Breckling (1989), in a major study of the wind speeds and directional aspects of weather patterns.

7.3 Spatial analysis

Two standard problems involving spatial analysis of directional data were described in the Introduction to this chapter. At present, the methods available for tackling, say, the one described in Example 7.3 are largely of an exploratory nature, relying on graphical techniques such as density estimation (§**2.2**(*iv*)) and spatial smoothing. The spatial smoothing techniques available are just two-dimensional versions of those described in §**7.2.2** and will not be elaborated further here. Instead, we shall illustrate the effect of one of the smoothing techniques (*loess*), and also describe briefly a general way of proceeding to identify so-called *homogeneous structural domains*.

(i) Spatial smoothing. With spatial smoothing (or spatial filtering as it is sometimes called) a general smoothing algorithm is applied to some subset of the data selected by specifying an arc, or angular window, of orientations within which the data must lie, e.g. all data falling within the arc $0° - 30°$ (equivalently, for axial data, $180° - 210°$).

The raw data falling within this arc should be plotted anyway, to see what general features (such as gradual shift of mean trend, across the area under

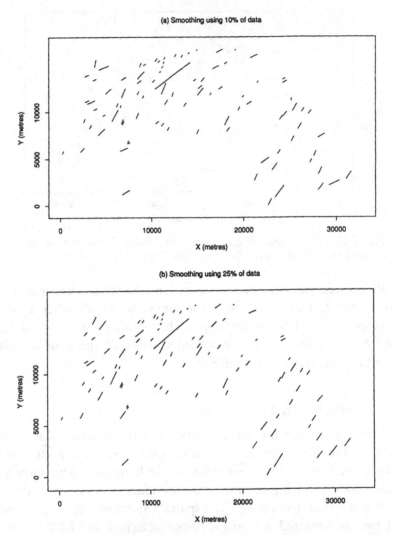

Fig. 7.14 Two smoothed versions of the subset of lineaments in Figure 7.12, using local regression with different spans. The more heavily smoothed graph suggests some degree of gradual migration of the lineament orientations across the region. (a) smoothing using 10% of data. (b) smoothing using 25% of data. See Example 7.9.

study) may be evident already. The objective of smoothing this subset is to emphasise such features if they are not sufficiently evident in the raw data plot.

If the whole data set is included in the arc, then a different aspect is highlighted, namely the predominant trends in each part of the region.

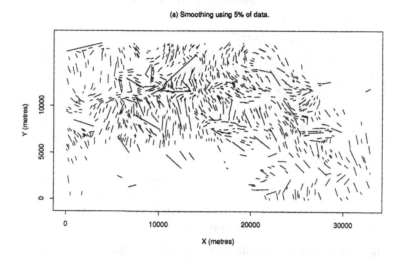

(a) Smoothing using 5% of data.

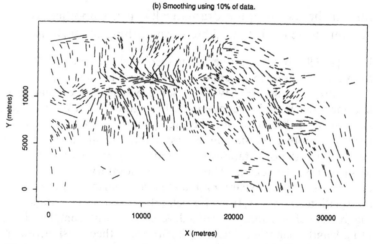

(b) Smoothing using 10% of data.

Fig. 7.15 Three smoothed versions of the lineaments in Figure 7.3, using local regression with different spans. Different degrees of smoothing bring out different aspect of the pattern. (a) Smoothing using 5% of data. (b) Smoothing using 10% of data. (c) Smoothing using 25% of data (*see next page*). See Example 7.9.

One aspect worth highlighting is the possible need to weight the directions or axes differentially in the smoothing process, according to their lengths. This is particularly so for the data in Figure 7.3, where there is a considerable range in lineament lengths. If this variation is not accommodated, major

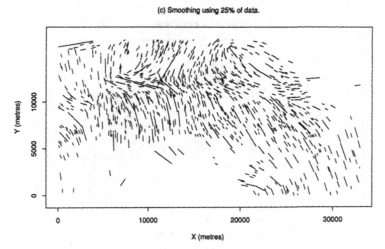

Fig. 7.15 (*cont.*) (c) Smoothing using 25% of data. See Example 7.9.

lineaments can be swung quite dramatically in a smoothed version of the data. Weighting is generally available in smoothing algorithms.

Example 7.9 For the lineament data displayed in Figure 7.3, it is appropriate to examine subsets in different arcs. Figure 7.13 shows the data in the 45° arc centred on the NE–SW axis. Some patches are evident in which this lineament set is under-represented. In Figure 7.14, two smoothed version are displayed, based on *loess* using spans of 10% and 25%. The more heavily smoothed version in Figure 7.14(b) seems to be revealing some tendency for the orientations to exhibit some systematic rotation from site to site across the region. This sort of exploratory analysis can be carried out for a set of arcs spanning the complete range, and the various bits of information combined. It is also helpful to explore the data set as a whole. Figure 7.15 shows three smoothed versions of the complete lineament set using length-weighting. The most lightly smoothed version (Figure 7.15(a)) clarifies the patterns locally, with the shortest lineaments altered slightly so that they conform more closely to their neighbours; the most heavily smoothed (Figure 7.15(c)) brings out gross predominant trends, although at the cost of rotating at least one major lineament substantially, so that the result may be misleading. Figure 7.15(b) preserves some local structure as well as revealing larger-scale patterns. The next step would be to relate possible rotations to the known geological history of the region.

References and footnotes. The spline method of Watson (1985) (see also Mendoza 1986) is an alternative form of smoothing. A general discussion of the spline method and methods based on local regression can be found in Hansen & Mount (1990), in the context of smoothing crustal

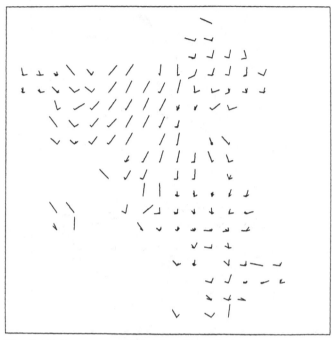

Fig. 7.16 Summary of fracture data from a colliery, after the data have been grouped into grid cells, then summarised in terms of the directions and strengths of the modal groups in each cell.

stress orientation measurements. Young (1987a, 1987b) has used a kriging approach to spatial smoothing of unit vectors and axes.

(ii) **Spatial clustering.** As a simple example of the sort of problem we have in mind, consider a region which has been affected by three significant fracturing events. The first and second of these may have created two ubiquitous sets of fractures trending N–S and E–W, say. Subsequently, a relatively narrow strip which passes through the centre of the region is affected by the third fracturing event trending NE–SW. The region is thus divided into three *structural domains*, within each of which the fracture pattern is homogeneous. Two major objectives of domain analysis are to identify the number of homogeneous structural domains and to determine which are similar.

One way of identifying spatially non-contiguous areas of similar fracture pattern is to study the directional aspect of the data separately from the spatial aspect, and then to put any resulting clusters of directions back into their spatial context. As an example of how this might be done, the

Fig. 7.17 A display resulting from performing a Principal Coordinates'
Analysis (PCA) on the summary cell diagrams, or 'birds' feet', in Figure 7.17,
using the similarity measure $S(P, Q)$. Each bird's foot has been plotted out
in the $x-y$ space defined by the first two principal coordinates.

analysis of Fisher *et al.* (1985) will be described briefly: for fuller details,
the reference should be consulted. (The data set considered in that paper
consisted of fractures in the tunnel roof throughout a coal mine, with the
fractures tending to be of comparable length. By way of comparison, the
fracture data in Example 7.3 have a wide range of fracture lengths; however,
for the grid-based method below, this difference is less pronounced than it
might be for other methods.)

There are three basic steps to the procedure:

(1) Summarisation. A grid is placed on the data map. For each grid cell,
the trends of all fractures intersecting that grid cell are recorded. Denote
the number of non-empty grid cells by N. The sample of measurements
in each cell is then analysed to identify the number of modal groups, and
the relative size and frequency of each. For a typical cell in which, say,
m modal groups have been identified, denote the summary information by
$\{(\bar{\theta}_i, p_i), i = 1, \ldots, m\}$, where p_i is the proportion of the cell data in the modal
group with mean trend $\bar{\theta}_i$. A simple summary plot for this cell is then

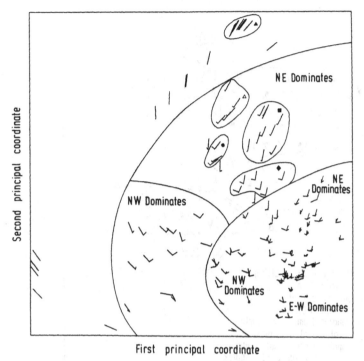

Fig. 7.18 A preliminary grouping of the birds' feet in Figure 7.17, by a geologist.

furnished by a plot of the m mean trends $\bar{\theta}_i$ as vectors of length p_i radiating from some common origin. The complete data set can now be displayed in summary form by plotting these N individual 'birds' feet' at the centres of their respective grid cells. Such a graph for the colliery data in Fisher *et al.* (1985) is shown in Figure 7.16.

(2) Clustering directional summaries based on angular similarity. It is not a straightforward matter to decide whether two birds' feet are similar, given that they might well have different numbers of trend vectors of varying lengths. A lengthy discussion of this point, together with details of an experiment to evaluate a number of possible measures, is given in the reference. The following *ad hoc* method (including formulae (7.28)–(7.31)) was found to give the best results in a test experiment.

Let $P = \{(\bar{\theta}_i, p_i), i = 1, \dots, m_1\}$ and $Q = \{(\bar{\phi}_i, q_i), i = 1, \dots, m_2\}$ be two summary sets, where $m_1 \le m_2$, say, and consider all possible match-ups of the m_1 mean trends $\bar{\theta}_1, \dots, \bar{\theta}_{m_1}$ with m_1-subsets of $\bar{\phi}_1, \dots, \bar{\phi}_{m_2}$. We shall define as *optimal* the particular match-up

$$(\bar{\theta}_1, \bar{\phi}_{j_1}), \dots, (\bar{\theta}_{m_1}, \bar{\phi}_{j_{m_1}})$$

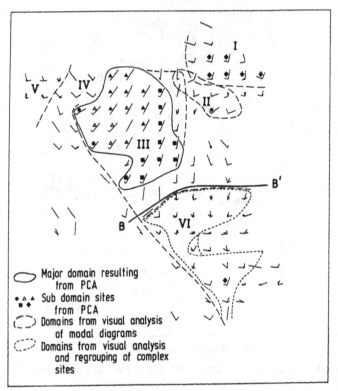

Fig. 7.19 Final domain analysis of the original colliery data, based on the groupings in Figure 7.18, further PCA, and other information.

which maximises the similarity

$$S_0(P, Q) = \frac{1}{2} \sum_{i=1}^{m_1} A_i B_i C_i \tag{7.28}$$

where

$$A_i = p_i + q_{j_i} \tag{7.29}$$
$$B_i = 1 - |\bar{\theta}_i - \bar{\phi}_{j_i}|/90 \tag{7.30}$$
$$C_i = 1 - 0.45|p_i - q_{j_i}| \tag{7.31}$$

and $\bar{\theta}_i - \bar{\phi}_{j_i}$ is the smaller of the two angular differences between the two mean trends (in degrees). The factor B_i measures similarity between the two means, the factor C_i similarity between the corresponding relative proportions (although the maximum effect of a gross dissimilarity between proportions is limited) and the factor A_i weights the product of these two similarities according to the total proportion their respective modes constitute of the entire samples. Finally, define the modified similarity

$$S(P,Q) = -\frac{1}{2}(1 - S_0(P,Q))^2 \qquad (7.32)$$

With all $\binom{N}{2}$ similarities between birds' feet calculated, carry out a principal coordinates' analysis (PCA – Gower 1966). In the $x-y$ space determined by the first two principal coordinates, plot each bird's foot P at the point given by its two principal coordinate values. This gives a two-dimensional clustering of the grid cells based on angular similarity alone, and ignoring spatial relationship.

The result of this procedure for the colliery data is shown in Figure 7.17. On this basis, several groups can be identified, as shown in Figure 7.18.

(3) Synthesis of directional and spatial similarities. The final stage is purely graphical: transfer the information about directional clustering back to the grid (e.g. by labelling the members of each cluster with a distinctive colour or symbol, or using linked PCA plot and grid views and highlighting on a computer screen). The result of this is shown in Figure 7.19.

Further details about this particular analysis and refinements of the method can be found in the reference. The main point to note is somewhat paradoxical: to perform the desired directional–spatial clustering, it may be necessary to decouple the directional from the spatial information initially.

8

Some modern statistical techniques for testing and estimation

8.1 Introduction

This chapter contains descriptions of two general classes of statistical methods used extensively in the earlier chapters of the book, namely *bootstrap methods* and *randomisation methods*.

The phrase 'bootstrap methods' refers to a class of computer-intensive statistical procedures which can often be helpful for carrying out a statistical test or for assessing the variability of a point estimate in situations where more usual statistical procedures are not valid and/or not available (e.g. the sampling distribution of a statistic is not known). Typically, but not always, these situations arise when only small amounts of data are available for analysis. The bootstrap methods not only provide satisfactory results in small samples, but compare favourably with other large-sample methods in larger samples for which the latter may be applicable. Accordingly, if one is in some doubt about the precise sample size at which to switch from a small-sample method to a large-sample method, both approaches can be used and the results compared. Generally, in estimation problems the large-sample methods provide a more convenient summary of the data, and are far simpler to implement. See Efron (1982) and Efron & Gong (1983) for accessible introductions to the subject. Hall (1992) has given a theoretical account of many aspects of the bootstrap.

In Section 8.2, the basic elements of the bootstrap method are described, in the context of a simple example for linear data. Section 8.3 sets out some general algorithms for bootstrap estimation, for circular data problems, and Section 8.4 provides corresponding algorithms for bootstrap testing.

Randomisation tests (equivalently permutation tests) are procedures which can be used to test certain statistical hypotheses with fairly minimal assumptions made about the form of probability distribution from which the data have been drawn. They are described in Section 8.5, through their application to two simple linear-data problems. As we shall have recourse to their use on a number of occasions, some explanation of their motivation is given, largely by illustrating their use with the two linear-data problems. Randomisation methods have been around for a long time, but are enjoying a new lease of life because of the ready availability of computational power.

8.2 Bootstrap methods for confidence intervals and hypothesis tests: general description

The essential features of the bootstrap approach are illustrated by the following trivial example for real (linear) data. Suppose we wish to estimate the mean μ of a linear distribution from a sample of 12 data points

$$19.2 \quad 16.2 \quad 10.7 \quad 16.6 \quad 3.6 \quad 18.1 \quad 8.6 \quad 15.3 \quad 14.0 \quad 14.2 \quad 16.9 \quad 13.4$$

and that we want an interval estimate, or confidence interval for μ. The sample mean of these data is $\bar{X} = 13.90$.

Stage 1: Re-sampling. Draw a random sample of size 10, *with replacement*, from the original sample (see details below), e.g.

$$3.6 \quad 19.2 \quad 8.6 \quad 15.3 \quad 16.9 \quad 8.6 \quad 3.6 \quad 14.2 \quad 10.7 \quad 10.7 \quad 10.7 \quad 16.9$$

This is the first *bootstrap* sample. Note that some of the original sample values occur more than once, and others not at all.

Stage 2: Bootstrap Estimate. Calculate the sample mean for the bootstrap sample:

$$\bar{X}_1^* = 11.58$$

Stage 3: Repetition. Carry out *Stages 1* and *2* a large number B (say 200) times, to obtain 200 bootstrap estimates $\bar{X}_1^*, \ldots, \bar{X}_{200}^*$:

$$11.58, 13.44, 12.71, 12.81, 13.95, 13.23, 14.96, \ldots, 15.36, 13.92, 14.52$$

Stage 4: Confidence Interval. Sort the bootstrap estimates into increasing order, to get

$$10.52, 10.65, 10.69, 11.05, 11.06, 11.08, 11.08, \ldots$$

$$\ldots, 15.75, 15.78, 15.80, 15.82, 15.92, 16.11, 16.22$$

For a 95% confidence interval for μ, choose the sixth smallest and sixth largest bootstrap values:

(11.08, 15.78)

This illustrates the basic idea of the bootstrap: we would like to calculate the distribution of the estimate $\bar{X}$, but we don't know the underlying distribution from which the data $X_1, \ldots, X_n$ were drawn. So we assume that these data *are* the true underlying distribution and calculate the distribution of $\bar{X}$ on this assumption, by repeated re-sampling from the data. The method as described can be improved in many ways, some of which we describe below in comments on the individual Stages. Basically, the more information we can incorporate about the data, the more effective our procedure becomes. Samples with very few data (e.g. $n < 8$) require special care to obtain reliable results.

Comments on Stage 1 (Re-sampling) and Stage 3 (Repetition)
Sampling with replacement. Sampling with replacement from a set of numbers $X_1, \ldots, X_n$ is very easily done. Let $U_1, \ldots, U_n$ be a set of pseudo-random numbers from the uniform distribution $U[0, 1]$ (see §**3.3.1**) and let

$$m_i = \text{integer part of } n \ U_i + 1, \ i = 1, \ldots, n \qquad (8.1)$$

Then select the numbers $X_{m_1}, \ldots, X_{m_n}$. For example, for the $n = 12$ data values used earlier, the pseudo-random values for $U_1, \ldots, U_{12}$ lead to the values

5 1 7 8 11 7 5 10 3 3 3 11

for $m_1, \ldots, m_{12}$ and thence to the first resample listed above.

Balanced re-sampling. Balanced re-sampling is a more efficient way of effecting *Stage 1*, provided that storage of large arrays is not a problem. Suppose we decide to use B bootstrap samples (each of size n). Create a vector of length $B \times n$ consisting of the sequence $X_1, \ldots, X_n$ repeated B times:

$$\overbrace{X_1, \ldots, X_n,}^{1} \overbrace{X_1, \ldots, X_n,}^{2} \ldots, \overbrace{X_1, \ldots, X_n}^{B} \qquad (8.2)$$

Now apply a random permutation to this vector, obtaining

$$\overbrace{X_1^*, \ldots, X_n^*,}^{1} \overbrace{X_{n+1}^*, \ldots, X_{2n}^*,}^{2} \ldots, \overbrace{X_{Bn-n+1}^*, \ldots, X_{Bn}^*}^{B} \qquad (8.3)$$

The result is B bootstrap samples which altogether contain equal numbers of the original $X_1, \ldots, X_n$.

An algorithm for generating a random permutation of a set of numbers is given by Nijenhuis & Wilf (1978, p. 62).

Incorporating extra information into the resampling. In many situations, we have additional information about the data which can be built into the re-sampling stage to improve the efficiency of the overall statistical procedure. There are two specific types of information of interest to us, here illustrated for our linear data example.

(i) *Symmetry assumption.* The data $X_1, \ldots, X_n$ are drawn from a distribution which is symmetric about its mean. With this assumption, modify the re-sampling as follows. If the method in Comment 1 has been used, for each re-sampled datum X^* say, let U be a new pseudo-random uniform number on $[0, 1]$. If $U < 0.5$, replace X^* by $2\bar{X} - X^*$. If the method in Comment 2 is used, replace the B n-sequences given in (8.3) by the $\frac{1}{2}B$ $2n$-sequences

$$\overbrace{X_1, \ldots, X_n, 2\bar{X} - X_1, \ldots, 2\bar{X} - X_n}^{1}, \ldots, \overbrace{X_1, \ldots, X_n, 2\bar{X} - X_1, \ldots, 2\bar{X} - X_n}^{\frac{1}{2}B} \tag{8.4}$$

$$\equiv Y_1, \ldots, Y_n, Y_{n+1}, \ldots, Y_{2n}, \ldots, \quad Y_{Bn} \tag{8.5}$$

say. Now apply a random permutation to the Y-sequence and proceed as before

(ii) *Parametric assumption.* The data $X_1, \ldots, X_n$ are drawn from a distribution of known form, e.g. the normal $N(\mu, \zeta^2)$ distribution, where μ and ζ^2 are unknown. With this assumption, estimate μ and ζ^2 by $\hat{\mu}, \widehat{\zeta^2}$ from the data $X_1, \ldots, X_n$, and then obtain each bootstrap sample by simulation from the $N(\hat{\mu}, \widehat{\zeta^2})$ distribution. For linear data, this approach can be used, for example, when comparing the means of two small samples of data from Normal distributions with differing variances. In such a case, if the variances are unknown, the distributions of the commonly used test statistics are very complicated, and a bootstrap approach with parametric resampling offers a satisfactory solution (see e.g. Fisher & Hall, 1990b).

Handling very small samples. For sample sizes up to $n = 6$, and possibly up to $n = 9$ (depending on the computer being used), it is feasible to enumerate *all n^n possible* bootstrap samples from $X_1, \ldots, X_n$ rather than simulating samples from $X_1, \ldots, X_n$. Thus, to take the trivial case $n = 3$, we would enumerate the distinct sample types

$$X_1, X_1, X_1, \qquad X_2, X_2, X_2, \qquad X_3, X_3, X_3$$
$$X_1, X_1, X_2, \qquad X_1, X_1, X_3, \qquad X_2, X_2, X_1, \qquad X_2, X_2, X_3,$$
$$X_3, X_3, X_1, \qquad X_3, X_3, X_2,$$
$$X_1, X_2, X_3$$

and associate with each such configuration a *weight* or likelihood of occurrence. Thus, for the type of sample consisting of repetitions of a single value, like X_1, X_1, X_1, the weight would be $\frac{1}{27}$, whereas for X_1, X_2, X_3, the weight is $\frac{6}{27}$. Equivalently, we can think of there being 27 ($n^n, n = 3$) possible bootstrap samples, with X_1, X_1, X_1 appearing once and X_1, X_2, X_3 six times. (The total number of distinct sample types for a sample of size n is $\binom{2n-1}{n}$.) These aspects are discussed further in Fisher & Hall (1991a) and Hall (1992).

Handling ties in small samples. A common reason for the existence of ties is rounding. Since ties can cause some numerical difficulties in implementing the bootstrap method in small samples, it is suggested that a small random number, related to the size of the rounding error, be added to *each* resampled datum. For example, suppose the data have been rounded to the nearest 5 units, and we get the resampled value $X^* = 15$. Let U be a pseudo-random uniform number from $U[0, 1]$. Then $V = 5 \times (U - 0.5)$ is a pseudo-random value uniformly distributed on $(-2.5, 2.5)$. Use as the re-sampled value $X^{**} = X^* + V = 15 + V$.

Comments on Stage 2 (Bootstrap estimate)
Confidence regions based on 'pivotal' statistics. In the example we have given, the bootstrap confidence region has been calculated directly from the bootstrap estimates of the mean. However, substantially better results can be obtained by calculating an interval of specified probability content interval from a 'pivotal', or re-scaled version of $\bar{X}$, and then transforming this interval to a confidence interval for the desired quantity μ. For a bootstrap sample $X_1^*, \ldots, X_n^*$ with mean $\widehat{X}^*$, the pivotal quantity is

$$\bar{Y}^* = (\bar{X}^* - \bar{X})/S^* \tag{8.6}$$

where

$$s^* = \left\{ \sum (X_i^* - \bar{X}^*)^2 / n \right\}^{\frac{1}{2}} \tag{8.7}$$

General use of pivotal statistics. In many situations, the parameter of interest, ω say, is a function of averages. Correspondingly, the sample estimate $\widehat{\omega}$ may be a function of sample averages. In the example we have just looked at, this is trivial:

$$\hat{\omega}(X_1,\ldots,X_n) = \bar{X} = (X_1 + \cdots + X_n)/n \tag{8.8}$$

A slightly more complicated case occurs when $\omega = \zeta^2$, the population variance. Then

$$\hat{\omega}(X_1,\ldots,X_n) = \sum_{i=1}^{n} (X_i - \bar{X})^2/n \tag{8.9}$$

which is easily re-expressible as

$$\hat{\omega}(X_1,\ldots,X_n) = \sum_{i=1}^{n} X_i^2/n - \bar{X}^2 \tag{8.10}$$

$$= \bar{U} - \bar{X}^2, \quad \bar{U} = \sum_{i=1}^{n} X_i^2/n \tag{8.11}$$

So, $\hat{\omega}$ is a function of two means $\bar{U}$ and $\bar{X}$. If we can construct a confidence region for the average vector $(\bar{U}, \bar{X}) = \bar{\mathbf{Y}}$ say, then this region can be simply transformed to a confidence interval for ζ^2. Now recall the preceding remark about pivotal statistics: we can do better by using a confidence region based on a *pivotal* version of $(\bar{U}, \bar{X})$. (Such a pivotal might take the form $S^{-\frac{1}{2}}(\bar{Y} - \mu)$, where μ is the (vector) mean of $\bar{Y}$ and S is its associated sample covariance matrix.) We shall not proceed further with this particular example. The main point to note is that, for applications in this book, we will generally be able to employ this approach except for samples of very small size.

Comments on Stage 4

Relation to pivotal statistics.   In view of the comments on *Stage 2*, we will usually calculate confidence regions based on pivotal statistics, and then transform them.

Use of assumptions.   We may be prepared to assume that the underlying distribution is symmetric about its mean direction and, indeed, to have included this in our re-sampling (cf. Comment 3 and *Stage 1*). In this case, it makes sense to construct symmetric confidence intervals.

Display of complete bootstrap distribution.   Occasionally, we may wish to convey more information about the distribution of our estimate than is given by a single (bootstrap) confidence interval. This can be done by calculating a nonparametric density estimate from the bootstrap values $\hat{\theta}_1^*,\ldots,\hat{\theta}_B^*$ say (cf. §2.2(*iv*)). A graph of the density estimate then conveys a picture of the variability of the estimate: this is illustrated in Example 4.21. If estimating a mean direction, use the simple rule (2.4) to decide the amount of smoothing.

8.3 Bootstrap methods for circular data: confidence regions for the mean direction

8.3.1 Introduction

We shall describe the use of the bootstrap method in terms of a series of Stages, as was done in §**8.2**. Within each Stage, there may be a number of alternative techniques. An appropriate choice of technique for each Stage is suggested in the relevant section of the book. Application to two or more samples is discussed separately, in §**8.3.3**. Very small sample sizes present special difficulties for the bootstrap method; a way of coping with these, the so-called *iterated bootstrap*, is discussed in §**8.3.4**. Four basic numerical procedures (Algorithms 1–4) are described in §**8.3.5**.

8.3.2 Methods for a single sample

Denote the sample of data by $\theta_1, \ldots, \theta_n$. Calculate the 2×1 vector $\mathbf{Z}_0 = \mathbf{z}$ and the 2×2 matrix $\mathbf{U}_0 = \mathbf{u}$ from Algorithm 1 in §**8.3.5**, and the 2×2 matrix $\mathbf{V}_0 = \mathbf{v}$ from Algorithm 2.

Stage 1: Re-sampling

Technique 1: Basic method. Let $u_1, \ldots, u_n$ be pseudo-random numbers from the uniform $U[0, 1]$ distribution (cf. §**3.3.1**) and calculate

$$i_1 = \text{integral part of } nu_1 + 1$$

$$\vdots$$

$$i_n = \text{integral part of } nu_n + 1 \tag{8.12}$$

Form the bootstrap sample

$$\theta_1^* = \theta_{i_1}, \ldots, \theta_n^* = \theta_{i_n} \tag{8.13}$$

Note: if storage is not a problem on the computer, a better approach is to generate *all* the required bootstrap data in one operation, using balanced resampling as described in Comment 2 on *Stage 1* in §**8.2**.

Technique 2: Symmetric distribution. Use Technique 1 to obtain $\theta_1^* = \theta_{i_1}, \ldots, \theta_n^* = \theta_{i_n}$. Let $u_1', \ldots, u_n'$ be a new set of pseudo-random uniform $[0,1]$ numbers. For each i, $i = 1, \ldots, n$, if $u_i' < 0.5$ replace θ_i^* by $2\bar{\theta} - \theta_i^*$

in the bootstrap sample, where $\bar{\theta}$ is the sample mean direction of $\theta_1, \ldots, \theta_n$.

Technique 3: Parametric bootstrap, e.g. von Mises distribution. Let $\hat{\mu}$ and $\hat{\kappa}$ be the usual estimates of μ and κ for the von Mises distribution (2.9) and (4.41). Simulate n values $\theta_1^*, \ldots, \theta_n^*$ from $VM(\hat{\mu}, \hat{\kappa})$ using the algorithm in §3.3.6.

Stage 2: Bootstrap Estimate. For the bootstrap sample $\theta_1^*, \ldots, \theta_n^*$, calculate the vector $\mathbf{Z}_1$ (say) $= \mathbf{z}$ from Algorithm 1 in §8.3.5, the 2×2 matrix $\mathbf{W}_1$ (say) $= \mathbf{w}$ from Algorithm 3 in §8.3.5, and finally, the bootstrap estimate $\hat{\mu}_B$ of the mean direction for this bootstrap sample, from Algorithm 4 in §8.3.5. Label this estimate $\hat{\mu}_1^*$.

Stage 3: Repetition. Use *Stages 1* and *2* to obtain a total of B (e.g. 200) bootstrap estimates $\hat{\mu}_1^*, \ldots, \hat{\mu}_B^*$ of the mean direction.

Stage 4: Confidence Interval

Technique 1: Basic method. To get a $100(1 - \alpha)\%$ confidence interval for the unknown mean direction, calculate

$$\gamma_b = \hat{\mu}_b^* - \hat{\theta}, (-\pi \leq \gamma_b < \pi), \quad b = 1, \ldots, B \tag{8.14}$$

and sort into increasing order, to obtain $\gamma_{(1)} \leq \cdots \leq \gamma_{(B)}$ say. Let $l = $ integer part of $(\frac{1}{2} B\alpha + \frac{1}{2})$ and $m = B - l$. The confidence interval for μ is $(\bar{\theta} + \gamma_{(l+1)}, \bar{\theta} + \gamma_{(m)})$.

(In essence, we are counting in $l + 1$ values from each extreme of the bootstrap set of estimates $\hat{\mu}_1^*, \ldots, \hat{\mu}_B^*$ relative to $\bar{\theta}$.) See also the method in *Technique 3*.

Technique 2: Symmetric distribution. To get a $100(1 - \alpha)\%$ confidence interval for the unknown mean direction, calculate $\psi_b = |\hat{\mu}_b^* - \bar{\theta}|, b = 1, \ldots, B$, and sort into increasing order to obtain $\psi_{(1)} \leq \cdots \leq \psi_{(B)}$ say. Let $l = $ integer part of $B\alpha + \frac{1}{2}, m = B - l$. The confidence interval for μ is $\bar{\theta} \pm \psi(m)$. See also the method in *Technique 3*.

Technique 3: Display of bootstrap distribution. The whole bootstrap distribution can be displayed by fitting a nonparametric density estimate to the values $\hat{\mu}_1^*, \ldots, \hat{\mu}_B^*$, as described in §2.2(iv), *Steps 1, 2* and *3*. If the assumption has been made that the underlying distribution is symmetric about its mean direction, include the corresponding modification for density estimation.

8.3.3 *Handling very small samples*

One reason for using pivotal methods to calculate bootstrap confidence regions rather than the simple method described in §**8.2** is that pivotal methods tend to give confidence regions which are closer to the nominal confidence level. What is meant by '95% confidence level'? It means that, if we continue to calculate 95% confidence intervals for the mean direction from sample after sample, in the long run about 95% of these intervals will actually contain the true unknown value. However, in many situations, we will aim at 95% intervals (the *nominal coverage*) yet, because of the approximate nature of our calculations or the approximate rather than exact validity of our assumptions, we may only achieve 92% intervals (the *actual coverage*). Bootstrap confidence intervals rarely give actual coverage equal to nominal coverage, although the difference is negligible in many cases. Use of confidence intervals based on pivotal statistics tends to assist in this. Further, when the pivotal-based methods are in error, they tend to be conservative, that is, actual coverage exceeds nominal coverage.

Very small sample sizes – n smaller than 7 or 8 – tend to produce poor coverages for bootstrap confidence intervals, regardless of which method is used to produce them, and pose particular problems for pivotal methods because of difficulties in estimating covariance matrices from resamples containing a high degree of replication of data values. However, if the simple bootstrap method is used (the analogue of §**8.2**), poor coverage will result anyway. One way to improve this is to attempt to calibrate the bootstrap method by estimating the relationship between nominal coverage and actual coverage and then selecting a nominal coverage which will produce the actual coverage we want. Again, this can only be done very approximately for small samples, but considerable improvements can result. The method is particularly effective in the context of the parametric bootstrap (§**8.3.2**). Ultimately, of course, there is no substitute for collecting larger samples.

The calibration technique is known as the *iterated bootstrap* and involves doing a second bootstrap experiment within the basic bootstrap experiment. As such, it is rather more computationally expensive than the basic bootstrap, although this is not a major consideration in small samples. It is described here in skeletal form in the context of calculating a $100(1 - \alpha)\%$ confidence interval for the mean direction μ, based on a random sample of data $\theta_1, \ldots, \theta_n$ with sample mean direction $\hat{\mu} = \bar{\theta}$: the various enhancements set out in §**8.3.2** *Stage 1* should still be used where possible.

Stage 0: Initialisation of probabilities. Set

$$p_1 = p_2 = \cdots = p_B = 0 \tag{8.15}$$

and

$$q_1 = q_2 = \cdots = q_B = 0 \tag{8.16}$$

Stage 1: Re-sampling. Draw a random sample of size n, $\theta_1^*, \ldots, \theta_n^*$ with replacement from $\theta_1, \ldots, \theta_n$.

Stage 2: Bootstrap Estimate. Calculate the bootstrap estimate of mean direction $\hat{\mu}_1^*$ from this resample (cf. (2.9)). Also carry out the following iterated bootstrap procedure, which yields a confidence interval for $\hat{\mu}$ (*not* for μ):

> **Stage 2.1: Re-sampling.** Draw a random sample of size n, $\theta_1^{**}, \ldots, \theta_n^{**}$, with replacement from $\theta_1^*, \ldots, \theta_n^*$.
>
> **Stage 2.2: Bootstrap Estimate.** Calculate the iterated bootstrap estimate $\hat{\mu}_1^{**}$ from this resample.
>
> **Stage 2.3: Repetition.** Use *Stages 2.1* and *2.2* to obtain a total of B independent bootstrap estimates $\hat{\mu}_1^{**}, \ldots, \hat{\mu}_B^{**}$.
>
> **Stage 2.4: Cumulate probabilities.** Calculate
>
> $$p_0 = 0, \quad p_j = p_{j-1} + (1/B)I\,[\hat{\mu} < \hat{\mu}_j^{**}], \quad j = 1, \ldots, B \tag{8.17}$$
>
> and
>
> $$q_0 = 1, \quad q_j = q_{j-1} - (1/B)I\,[\hat{\mu} > \hat{\mu}_j^{**}], \quad j = 1, \ldots, B \tag{8.18}$$
>
> where the indicator function $I\,[A]$ is 1 if the event A is true, otherwise 0.

Stage 3: Repetition. Use *Stages 1* and *2* to obtain a total of B (e.g. 200) bootstrap estimates $\hat{\mu}_1^*, \ldots, \hat{\mu}_B^*$ of the mean direction, and final sets of probabilities $p_1, \ldots, p_B$ and $q_1, \ldots, q_B$.

Stage 4: Confidence Interval. Arrange the bootstrap estimates of mean direction into increasing order to get

$$\hat{\mu}_{(1)}^* \leq \cdots \leq \hat{\mu}_{(B)}^* \tag{8.19}$$

For a $100(1 - \alpha)\%$ confidence interval for the unknown mean direction, find the integer J corresponding to the largest p_j less than or equal to $\frac{1}{2}\alpha$, and similarly the integer K corresponding to the largest q_k less than or equal to $\frac{1}{2}\alpha$. Then the $100(1-\alpha)\%$ bootstrap confidence interval is given by

$$(\widehat{\mu}^*_{(J+1)}, \widehat{\mu}^*_{(K-1)}) \tag{8.20}$$

To calculate a nonparametric density estimate (cf. §2.2(*iv*)) from the bootstrap values $\widehat{\mu}^*_1, \ldots, \widehat{\mu}^*_B$, let

$$P_j = \tfrac{1}{2}(p_{j+1} - p_j + q_{j-1} - q_j), \quad j = 1, \ldots, B \tag{8.21}$$

(where $p_{B+1} = q_0 = 1$) and instead of (2.4) use the formula

$$\widehat{f}(\theta) = h^{-1} \sum_{i=1}^{n} P_i w\left\{(\theta - \theta_i)/h\right\} \tag{8.22}$$

8.3.4 Methods for combining two or more samples

Little extra is required to extend the bootstrap methods for a single sample to two or more samples. Suppose that we have r (≥ 2) random samples of circular data

$$\theta_{i1}, \ldots, \theta_{in_i}, \quad i = 1, \ldots, r$$

from unimodal distributions which are believed to have a common mean direction μ, and that we wish to form a pooled estimate of μ. As can be seen from §5.3.5, the manner in which the information about μ is combined depends on factors such as differing sample sizes and dispersions. Given that a suitable method has been selected, proceed as follows. First, calculate the estimate $\widehat{\mu}$ say, using this method. Then carry out the bootstrap algorithm to obtain a confidence region for μ.

Stage 1: Re-sampling. For each sample i, $i = 1, \ldots, r$, draw a resample $\theta^*_{i1}, \ldots, \theta^*_{in_i}$ from $\theta_{i1}, \ldots, \theta_{in_i}$.

Stage 2: Bootstrap estimate of μ. Calculate the estimate μ^* say, for this set of r resamples, in the same way that $\widehat{\mu}$ was calculated.

Stages 3 and 4 are then as for §8.3.2, yielding a confidence region for μ. For very small sample sizes, the methods of §8.3.3 can be used.

8.3.5 Computing algorithms

Let $\phi_1, \ldots, \phi_n$ be an arbitrary sample of directions (e.g. they might be the original sample, or one of the bootstrap samples), and let

$$x_i = \cos \phi_i, \quad y_i = \sin \phi_i, \quad i = 1, \ldots, n \tag{8.23}$$

The algorithms are written in terms of the values $(x_i, y_i), i = 1, \ldots, n$.

Algorithm 1. Mean vector (**z**) and covariance matrix (**u**) of $(x_i, y_i), i = 1, \ldots, n$.

$$z_1 = \sum_{i=1}^{n} x_i/n, \quad z_2 = \sum_{i=1}^{n} y_i/n, \quad \mathbf{z} = \begin{pmatrix} z_1 \\ z_2 \end{pmatrix} \tag{8.24}$$

$$u_{11} = \sum_{i=1}^{n} (x_i - z_1)^2/n, \quad u_{22} = \sum_{i=1}^{n} (y_i - z_2)^2/n \tag{8.25}$$

$$u_{12} = u_{21} = \sum_{i=1}^{n} (x_i - z_1)(y_i - z_2)/n, \quad \mathbf{u} = \begin{pmatrix} u_{11} & u_{12} \\ u_{21} & u_{22} \end{pmatrix} \tag{8.26}$$

Algorithm 2. Square root (**v**) of a positive definite symmetric 2 x 2 matrix **u**.

$$\beta = (u_{11} - u_{22})/(2u_{12}) - [(u_{11} - u_{22})^2/(4u_{12}^2) + 1]^{\frac{1}{2}} \tag{8.27}$$

$$t_1 = (\beta^2 u_{11} + 2\beta u_{12} + u_{22})^{\frac{1}{2}}/(1 + \beta^2)^{\frac{1}{2}} \tag{8.28}$$

$$t_2 = (u_{11} - 2\beta u_{12} + \beta^2 u_{22})^{\frac{1}{2}}/(1 + \beta^2)^{\frac{1}{2}} \tag{8.29}$$

$$v_{11} = (\beta^2 t_1 + t_2)/(1 + \beta^2), \quad v_{22} = (t_1 + \beta^2 t_2)/(1 + \beta^2) \tag{8.30}$$

$$v_{12} = v_{21} = \beta(t_1 - t_2)/(1 + \beta^2), \quad \mathbf{v} = \begin{pmatrix} v_{11} & v_{12} \\ v_{21} & v_{22} \end{pmatrix} \tag{8.31}$$

Algorithm 3. Inverse (**w**) of square root of a positive definite 2 × 2 matrix **u**.

$$\beta = (u_{11} - u_{22})/(2u_{12}) - [(u_{11} - u_{22})^2/(4u_{12}^2) + 1]^{\frac{1}{2}} \tag{8.32}$$

$$t_1 = (1 + \beta^2)^{\frac{1}{2}}/(\beta^2 u_{11} + 2\beta u_{12} + u_{22})^{\frac{1}{2}} \tag{8.33}$$

$$t_2 = (1 + \beta^2)^{\frac{1}{2}}/(u_{11} - 2\beta u_{12} + \beta^2 u_{22})^{\frac{1}{2}} \tag{8.34}$$

$$w_{11} = (\beta^2 t_1 + t_2)/(1 + \beta^2), w_{22} = (t_1 + \beta^2 t_2)/(1 + \beta^2) \tag{8.35}$$

$$w_{12} = w_{21} = \beta(t_1 - t_2)/(1 + \beta^2); \quad \mathbf{w} = \begin{pmatrix} w_{11} & w_{12} \\ w_{21} & w_{22} \end{pmatrix} \tag{8.36}$$

Algorithm 4. Estimate of mean direction.

Given the quantities z_0 and v_0 from the original sample, and w_B (say) and z_B from a bootstrap sample (e.g. $w_B = w_1$ and $z_B = z_1$, for the first bootstrap sample), calculate C_B and S_B, the cosine and sine respectively of the bootstrap estimate $\hat{\mu}_B$ of the mean direction from

$$\begin{pmatrix} c_B \\ s_B \end{pmatrix} = z_0 + v_0 w_B (z_B - z_0) \tag{8.37}$$

$$\begin{pmatrix} C_B \\ S_B \end{pmatrix} = (c_B^2 + s_B^2)^{-\frac{1}{2}} \begin{pmatrix} c_B \\ s_B \end{pmatrix} \tag{8.38}$$

$\hat{\mu}_B$ can now be determined from equations (2.8) and 2.9).

8.4 Bootstrap methods for circular data: hypothesis tests for mean directions

8.4.1 Introduction

The methods to be used are arranged in a series of Stages, similar in spirit to those described in §**8.2**. In some stages, there may be a number of alternative techniques, appropriate choice amongst which can be made by consulting the relevant Sections in the book. Application to two or more samples is discussed separately, in §**8.4.4**.

8.4.2 Methods for a single sample

We denote the sample of data by $\theta_1, \ldots, \theta_n$, the hypothesised value for the mean direction by μ_0, and the sample mean direction by $\bar{\theta}$. Let $t \equiv T(\theta_1, \ldots, \theta_n; \mu_0)$ be the value of the test statistic suggested for the *large-sample* test of the hypothesis $H_0 : \mu = \mu_0$. Also define the shifted sample of data

$$\psi_1 = \theta_1 - (\bar{\theta} - \mu_0), \ldots, \psi_n = \theta_n - (\bar{\theta} - \mu_0) \tag{8.39}$$

Under the null hypothesis, this shifted sample is centred at the hypothesised mean direction μ_0.

Stage 1: Re-sampling. This is very similar to what was described in §**8.3.2**, *Stage 1*, with the crucial difference that the resample is drawn from the *shifted* sample $\psi_1, \ldots, \psi_n$. Denote the bootstrap sample by $\theta_1^*, \ldots, \theta_n^*$.

Stage 2: Bootstrap Test Statistic. For the bootstrap sample $\theta_1^*, \ldots, \theta_n^*$ calculate

$$t_1^* = T(\theta_1^*, \ldots, \theta_n^*; \bar{\theta}) \tag{8.40}$$

Stage 3: Repetition. Use *Stages 1* and *2* to obtain a total of B (e.g. 200) values $t_1^*, \ldots, t_B^*$ of the test statistic.

Stage 4: Critical Region. We consider two-sided tests, i.e. those for which the null hypothesis (H_0) is that $\mu = \mu_0$, and the alternative (H_1) is that $\mu \neq \mu_0$. Re-arrange the values $|t_1^*|, \ldots, |t_B^*|$ into increasing order, to get $u_1 \leq \cdots \leq u_B$ say. For a test at level $100\alpha\%$, set $m_\alpha =$ integer part of $B(1-\alpha) + 1$; reject H_0 in favour of H_1 if $|t| \geq u_{m_\alpha}$. Alternatively, the *significance probability* of t is m/B, where m is an integer such that $u_m \leq |t| \leq u_{m+1}$.

(One-sided tests such as $H_0 : \mu \leq \mu_0$ versus $H_1 : \mu \geq \mu_0$ are rarely of interest, but can be carried out with concentrated data; see e.g. **§4.6(i)**.)

8.4.3 Handling very small samples

Refer to the preamble to **§8.3.3** (the corresponding section for the calculation of confidence intervals) for a discussion of when this method will be helpful.

Stage 1: Re-sampling. This is exactly as described in §8.3.2, *Stage 1*. Denote the bootstrap sample by $\theta_1^*, \ldots, \theta_n^*$.

Stage 2: Bootstrap Test Statistic. For the bootstrap sample $\theta_1^*, \ldots, \theta_2^*$, calculate $t_1 = T(\theta_1^*, \ldots, \theta_n^*; \bar{\theta})$.

> **Stage 2.1: Re-sampling.** Draw a random sample of size n, $\theta_1^{**}, \ldots, \theta_n^{**}$ with replacement from $\theta_1^*, \ldots, \theta_n^*$.
>
> **Stage 2.2: Iterated bootstrap test statistic.** For the iterated bootstrap sample, calculate
>
> $$t_1^{**} = T|(\theta_1^{**}, \ldots, \theta_n^{**}; \bar{\theta}^*)$$
>
> where θ^* is the sample mean direction of $\theta_1^*, \ldots, \theta_n^*$.
>
> **Stage 2.3: Repetition.** Use *Stages 2.1* and *2.2* to obtain a total of B (e.g. 200) values
>
> $$t_1^{**}, \ldots, t_B^{**}$$
>
> of the test statistic.

Stage 2.4: Iterated bootstrap significance probability. Re-arrange the values $| t_1^{**} |, \ldots, | t_B^{**} |$ into increasing order, to get $u_1^* \leq \cdots \leq u_B^*$ say. Then calculate the iterated bootstrap significance probability $P_1^* = m^*/B$, where m^* is an integer such that $u_{m^*}^* \leq | t | \leq u_{m^*+1}^*$.

Stage 3: Repetition. Use *Stages 1* and *2* to obtain a total of B values $t_1^*, \ldots, t_B^*$ of the test statistic, with corresponding values $P_1^*, \ldots, P_B^*$.

Stage 4: Bootstrap significance probability Re-arrange the values $| t_1^* |, \ldots, | t_B^* |$ into increasing order, to get $u_1 \leq \cdots \leq u_B$ say. Then calculate the *uncalibrated* bootstrap significance probability $P_0 = m/B$, where m is the largest integer such that $u_m \leq | t | \leq u_{m+1}$. To calibrate, or adjust, this raw value, sort the numbers $P_1^*, \ldots, P_B^*$ into increasing order,

$$P_{(1)}^* \leq \cdots \leq P_{(B)}^* \tag{8.41}$$

say, plot the points

$$(b/(B+1), P_{(b)}^*), \quad b = 1, \ldots, B \tag{8.42}$$

and join them up, interpolating linearly between them. Locate the raw significance probability P_0 on the y-axis and read across to the line then down to the x-axis to get the adjusted significance probability.

8.4.4 Methods for comparing two or more samples

As in §**8.3.4**, little extra is required to extend the bootstrap methods for a single sample to two or more samples. Here, we give a skeletal outline of how to test the hypothesis that the mean directions of r (≥ 2) circular distributions are equal, based on samples of data $\theta_{i1}, \ldots, \theta_{in_i}, i = 1, \ldots, r$.

Let S_r be the large-sample test statistic for testing this hypothesis. Calculate its value S for the data set being analysed. Then calculate the *centred* samples

$$\psi_{i1} = \theta_{i1} - \bar{\theta}_i, \ldots, \psi_{in_i} = \theta_{in_i} - \bar{\theta}_i, \quad i = 1, \ldots, r \tag{8.43}$$

where $\bar{\theta}_i$ is the mean direction of sample i.

Stage 1: Re-sampling. For each sample i, $i = 1, \ldots, r$, draw a resample $\theta_{i1}^*, \ldots, \theta_{in_i}^*$ from $\psi_{i1}, \ldots, \psi_{in_i}$.

Stage 2: Bootstrap Test Statistic. Calculate the value of S_r, S^* say, for this set of r resamples.

Stages 3 and 4 are then as for §8.4.2, yielding a significance probability for S. This can be calibrated as in §8.4.3 if the sample sizes are very small.

8.5 Randomisation, or permutation, tests

Randomisation methods have been part of the statistical literature for many decades, unlike bootstrap methods. However, like bootstrap methods, they generally require a substantial amount of computing power to carry out (except with very small sample sizes) so have only become feasible relatively recently. In terms of their usage in other parts of this book, it will suffice to describe them in the context of two simple examples.

Suppose we have two random samples of data as follows

Sample 1:   1.2   3.3   6.2   6.6   6.8   7.1

($n_1 = 6$)

Sample 2:   3.9   7.7   11.9   13.0

($n_2 = 4$)

and wish to test the hypothesis that they were drawn from the same distribution, against the alternative that they were drawn from distributions with the same shape but different locations. A well-known statistic for testing this hypothesis (for *normally distributed data*) is the t-statistic

$$t = |\bar{X}_1 - \bar{X}_2|/s \tag{8.44}$$

where $\bar{X}_1$ and $\bar{X}_2$ are the respective sample averages and S^2 is a weighted combination of the separate sample estimates of the common variance. If the assumption of normality of the data is reasonable, then the distribution of t is known (Student's t with $n_1 + n_2 - 2$ degrees of freedom, when the null hypothesis is true); the hypothesis of equality of means (equivalently, of distributions) is rejected if the value of the test statistic t is too large.

What if we are not prepared to assume that the data are normally distributed? Then we can use a randomisation method. Under the null hypothesis, all $n_1 + n_2 = 10$ data points were drawn from the same distribution, so the allocation of the values 1.2, 3.3, 6.2, 6.6, 6.8 and 7.1 to the first sample and 3.9, 7.7, 11.9 and 13.0 to the second sample was purely chance. Given that we actually obtained the 10 values 1.2,...,13.0, we could equally well have obtained

Sample 1:   3.3   3.9   6.8   7.1   11.9   13.0

Sample 2:   1.2   6.2   6.6   7.7

or

Sample 1:   3.3   6.2   6.6   7.1   7.7   11.9

Sample 2:   1.2   3.9   6.8   13.0

and so on. There are $\binom{n_1+n_2}{n_1}$ = 210 possible ways of allocating 10 points to two samples of sizes 6 and 4, *each of which is equally likely to occur under the null hypothesis*. So, the value we obtained for t for our actual data ($t = 1.235$) could equally well have been any of the other 209 values corresponding to the 209 other possible pairs of samples. (An algorithm for enumerating all possible ways of dividing a set of points into two distinct subsets of given sizes is given in Nijenhuis & Wilf 1978, p. 26.)

We may then ask: is the value we obtained for our *actual* pair of samples extreme, or unusual, considering all the $N = 210$ values which were possible? To answer this, we calculate all N possible t values and arrange them in increasing order, to get the *randomisation distribution*:

$$t_{(1)} \leq t_{(2)} \leq \cdots \leq t_{(N-1)} \leq t_{(N)}$$

For the data set in hand, these values are (when rounded to two decimal places),

0.01   0.01   0.01   0.01   0.01   0.02 $\cdots$ 1.24   1.27   1.31

1.49   1.54   1.57   1.61   1.64   1.72   1.73   1.78   1.78   1.82

1.89   1.90   1.97   2.02   2.07

If we want a significance test at level $100\alpha\%$, let

$$l_\alpha = \text{integer part of } N\alpha + \frac{1}{2}, \quad m_\alpha = N - l_\alpha + 1 \tag{8.45}$$

Reject the null hypothesis of equality of means if

$$t \geq t_{(m)} \tag{8.46}$$

Alternatively, suppose that in the randomisation distribution, $t = t_{(m)}$, and set

$$p = \frac{N - m + 1}{N} \tag{8.47}$$

Then p is the *significance probability* of t. In the example, $m = 193$, so the significance probability of t is 0.086. There is a modest amount of evidence that the samples were drawn from populations with different means.

In the event that there are two or more tied values equal to the actual sample value t, let m_0 denote the number of tied values, and calculate the significance probability as

$$p = \frac{N-m}{N} + \frac{m_0}{2} \tag{8.48}$$

If at least one sample size is large, it will not be feasible to enumerate all possible $N = \binom{n_1+n_2}{n_1}$ allocations of the data points to two samples of sizes n_1 and n_2. In such cases, a large number (e.g. $N = 1000$) of *random* allocations of n_1 points to the first sample and the remaining points to the second sample can be generated using, for example, the algorithm in Nijenhuis & Wilf (1978, p. 39), and then proceed as above to generate an estimated randomisation distribution with these N points in it.

A second, rather different type of problem arises in the context of calculating the correlation of a circular random variable with another random variable (either linear or circular), which is treated in Chapter 6. Here, we shall demonstrate the basic principle in the context of a simple linear data problem.

Suppose we are interested in the extent to which two random variables X and Y are linearly related, and we have available n pairs of independent measurements of them, $(X_1, Y_1), \ldots (X_n, Y_n)$. A common measure of the linear correlation ρ between X and Y is the sample correlation coefficient

$$r = \frac{\sum_{i=1}^{n}(X_i - \bar{X})(Y_i - \bar{Y})}{[\sum_{i=1}^{n}(X_i - \bar{X})^2 \sum_{i=1}^{n}(Y_i - \bar{Y})^2]^{\frac{1}{2}}}$$

$$= \frac{\sum_{i=1}^{n} X_i Y_i - n\bar{X}\bar{Y}}{[\sum_{i=1}^{n}(X_i - \bar{X})^2 \sum_{i=1}^{n}(Y_i - \bar{Y})^2]^{\frac{1}{2}}} \tag{8.49}$$

where $\bar{X} = \sum_{i=1}^{n} X_i/n$ and $\bar{Y} = \sum_{i=1}^{n} Y_i/n$. Values of r near 0 correspond to little evidence of linear dependence.

Suppose, for example, that we have obtained the following sample of size 12

X:	1.7	2.7	7.3	5.1	2.3	6.2	1.8	3.2	9.6	1.6	5.8	−0.4
Y:	5.4	9.3	8.3	1.4	−0.7	1.0	−0.2	5.3	−0.0	4.6	−0.3	7.4

and wish to test the null hypothesis of zero correlation against the alternative hypothesis of negative correlation. (A plot of the data is shown in Figure 8.1.) We need to be able to assess whether the sample value r is 'significantly large and negative' given the null hypothesis that $\rho = 0$. We can do this by finding the randomisation distribution of r.

Suppose in fact that X and Y are uncorrelated. Then, *given the data values we actually obtained*, the pairing of 1.7 with 5.4, ..., −0.4 with 7.4 was

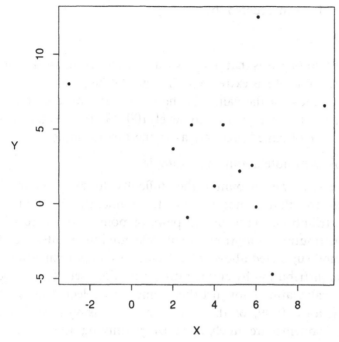

Fig. 8.1 Scatterplot of synthetic data. A randomisation test can be used to decide whether there is some evidence of negative correlation between X and Y.

entirely fortuitous: we could equally well have observed the pairing

X:	1.7	2.7	7.3	5.1	2.3	6.2	1.8	3.2	9.6	1.6	5.8	−0.4
Y:	9.3	8.3	1.4	−0.7	1.0	−0.2	5.3	−0.0	4.6	−0.3	7.4	5.4

and so on. In fact, in addition to the actual pairings we obtained in the sample, there are $N = n! - 1$ other possible ways of pairing n Xs with n Ys which could equally well have occurred instead. (Here, N still has the same meaning as in the first example, namely the number of points in the randomisation distribution, although it has a different mathematical formula.) For our example, this means that there are a total of $12! = 479\,001\,600$ equally likely sets of sample pairings given the two sets of values

$$\{1.7, 2.7, 7.3, 5.1, 2.3, 6.2, 1.8, 3.2, 9.6, 1.6, 5.8, -0.4\}$$

and

$$\{5.4, 9.3, 8.3, 1.4, -0.7, 1.0, -0.2, 5.3, -0.0, 4.6, -0.3, 7.4\}$$

In principle, then, we have to enumerate all these sample pairings (see e.g. Nijenhuis & Wilf, 1978, p. 54) and calculate the value of r for each. Denote the values so obtained, when ordered, by

$$r_{(1)} \leq r_{(2)} \leq \cdots \leq r_{(N)}$$

Since the alternative hypothesis is that $\rho < 0$, we are interested in ascertaining whether the observed value of r is extreme in the sense of being too large and negative, i.e. less than most of the values in the randomisation distribution. A test of significance at some prescribed level $100\alpha\%$, or a significance probability for r, can be obtained precisely as in the first example.

There are two points to note about this example:

• With this sample size, we encounter the difficulty touched on in the first example, namely, that is not feasible to enumerate the complete randomisation distribution. With just 12 pairs of points, the correlation randomisation distribution comprises nearly 500 million points. So we shall use the method suggested above, of calculating an approximation to the randomisation distribution by enumerating a random set of $N = 1000$ permutations and calculating r for just these randomly selected samples.

• The mathematical form (8.49) of the statistic we are using is such that several of its components are unchanged by permuting the Y-values, namely any part of the formula which is a function of just the Xs or just the Ys. In fact, the only part of the formula which has the Xs and Ys inextricably intertwined is $\sum_{i=1}^{n} X_i Y_i$, and r is a monotonic function of $\sum_{i=1}^{n} X_i Y_i$; that is, if $\sum_{i=1}^{n} X_i Y_i$ increases so does r, and if $\sum_{i=1}^{n} X_i Y_i$ decreases so does r. So, it suffices to base the randomisation test on the quantity $\sum_{i=1}^{n} X_i Y_i$, thereby reducing the amount of calculation considerably.

For the data in hand, $\sum_{i=1}^{n} X_i Y_i = 125.87$. Carrying out the calculations of $\sum_{i=1}^{n} X_i Y_i$ for 1000 randomly selected permutations results in an ordered set of randomisation values

70.67   71.75   71.93   79.34   80.09   80.25   80.64   82.88   83.34 $\cdots$

and the sample value is larger than the first 149 values in this sequence. Thus the significance probability of $\sum_{i=1}^{n} X_i Y_i$, or equivalently of r, is *estimated* as 0.15, which we would not expect to differ by much from the true value obtained from the complete randomisation distribution of N values. On the basis of this test, there is little evidence of negative correlation between X and Y.

Appendix A: Tables

A.1 Percentiles of the Normal $N(0, 1)$ distribution

This table gives values $x(\alpha)$ defined by the equation

$$\alpha = (2\pi)^{-\frac{1}{2}} \int_{x(\alpha)}^{\infty} \exp(-u^2) \, du$$

as shown in Figure A1.

If X is a Normal $N(0, 1)$ random variable, $\text{Prob}\{X \geq x(\alpha)\} = \alpha$. $x(\alpha)$ corresponds to the upper P percent point, where $P = 100x(\alpha)$. The lower P percent point $x(1 - \alpha)$ is obtained by symmetry, as $x(1 - \alpha) = -x(\alpha)$.

The two-tailed probability $\text{Prob}\{|X| \geq x(\alpha)\} = 2\alpha = 2P/100$.

α	$x(\alpha)$	α	$x(\alpha)$	α	$x(\alpha)$	α	$x(\alpha)$
0.50	0.0000	0.34	0.4120	0.18	0.9153	0.060	1.5551
0.49	0.0250	0.33	0.4395	0.17	0.9541	0.055	1.5985
0.48	0.0500	0.32	0.4673	0.16	0.9944	0.050	1.6452
0.47	0.0751	0.31	0.4954	0.15	1.0364	0.045	1.6958
0.46	0.1002	0.30	0.5240	0.14	1.0804	0.040	1.7511
0.45	0.1254	0.29	0.5530	0.13	1.1265	0.035	1.8123
0.44	0.1507	0.28	0.5825	0.12	1.1751	0.030	1.8812
0.43	0.1760	0.27	0.6125	0.11	1.2267	0.025	1.9604
0.42	0.2015	0.26	0.6430	0.10	1.2817	0.020	2.0542
0.41	0.2271	0.25	0.6742	0.095	1.3108	0.015	2.1705
0.40	0.2529	0.24	0.7060	0.090	1.3410	0.010	2.3268
0.39	0.2789	0.23	0.7386	0.085	1.3724	0.005	2.5762
0.38	0.3050	0.22	0.7720	0.080	1.4053	0.001	3.0905
0.37	0.3314	0.21	0.8062	0.075	1.4398	0.0005	3.2907
0.36	0.3580	0.20	0.8415	0.070	1.4761	0.0001	3.7191
0.35	0.3849	0.19	0.8778	0.065	1.5144	0.00005	3.8906

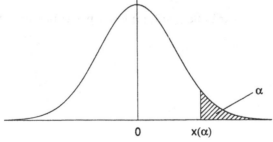

Fig. A.1 Plot of the equation giving rise to the values in Table A1.

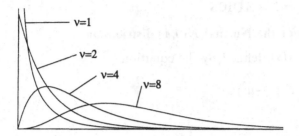

Fig. A.2 Various shapes of the χ^2-distribution for different v values.

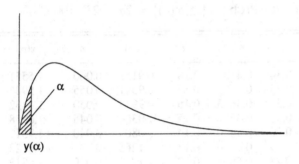

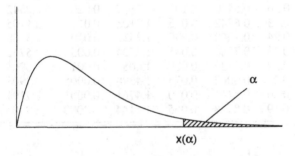

Fig. A.3 Plot of the χ^2-distribution with (a) lower percentiles and (b) upper percentiles shaped.

A.2 Percentiles of the χ_ν^2-distribution

These two tables give lower and upper percentiles for the χ^2-distribution with ν degrees of freedom (χ_ν^2). Figure A2 show the various shapes of the χ^2-distribution. For $\nu = 1$ and $\nu = 2$, the probability density function has its maximum at $x = 0$ and decreases to zero as x increases; for $\nu = 3, 4, \ldots$, the density is zero at $x = 0$, increasing to a maximum value then decreasing to zero.

(a) *Lower percentiles.* This table gives values $y(\alpha)$ defined by the equation

$$\text{Prob}\left\{\chi_\nu^2 \le y(\alpha)\right\} = \alpha$$

for selected values of the degrees of freedom ν and selected values of α as shown in Figure A3(a). For example, the lower 2% point of the χ_5^2 distribution is 0.752.

					α				
$y(\alpha)$	0.001	0.005	0.01	0.015	0.02	0.025	0.05	0.10	0.20
$\nu = 1$	0.00000	0.00004	0.00016	0.00035	0.00062	0.00098	0.0039	0.0157	0.0640
2	0.0020	0.0100	0.0201	0.0302	0.0404	0.0506	0.103	0.211	0.446
3	0.0243	0.0717	0.115	0.152	0.185	0.216	0.352	0.584	1.01
4	0.0908	0.207	0.297	0.368	0.429	0.484	0.711	1.06	1.65
5	0.210	0.412	0.554	0.662	0.752	0.831	1.15	1.61	2.34
6	0.381	0.676	0.872	1.02	1.13	1.24	1.64	2.20	3.07
7	0.598	0.989	1.24	1.44	1.56	1.69	2.17	2.83	3.82
8	0.857	1.34	1.65	1.86	2.03	2.18	2.73	3.49	4.59
9	1.15	1.73	2.09	2.33	2.53	2.70	3.33	4.17	5.38
10	1.48	2.16	2.56	2.84	3.06	3.25	3.94	4.87	6.18
11	1.83	2.60	3.05	3.36	3.61	3.82	4.57	5.58	6.99
12	2.21	3.07	3.57	3.91	4.18	4.40	5.23	6.30	7.81
13	2.62	3.57	4.11	4.48	4.77	5.01	5.89	7.04	8.63
14	3.04	4.07	4.66	5.06	5.37	5.63	6.57	7.79	9.47
15	3.48	4.60	5.23	5.65	5.98	6.26	7.26	8.55	10.3
16	3.94	5.14	5.81	6.26	6.61	6.91	7.96	9.31	11.2
17	4.42	5.70	6.41	6.88	7.26	7.56	8.67	10.1	12.0
18	4.90	6.26	7.01	7.52	7.91	8.23	9.39	10.9	12.9
19	5.41	6.84	7.63	8.16	8.57	8.91	10.1	11.7	13.7
20	5.92	7.43	8.26	8.81	9.24	9.59	10.9	12.4	14.6
21	6.45	8.03	8.90	9.47	9.91	10.3	11.6	13.2	15.4
22	6.98	8.64	9.54	10.1	10.6	11.0	12.3	14.0	16.3
23	7.53	9.26	10.2	10.8	11.3	11.7	13.1	14.9	17.2
24	8.08	9.89	10.9	11.5	12.0	12.4	13.9	15.7	18.1
25	8.65	10.5	11.5	12.2	12.7	13.1	14.6	16.5	18.9
26	9.22	11.2	12.2	12.9	13.4	13.8	15.4	17.3	19.8
27	9.80	11.8	12.9	13.6	14.1	14.6	16.2	18.1	20.7
28	10.4	12.5	13.6	14.3	14.8	15.3	16.9	18.9	21.6

					α				
$y(\alpha)$	0.001	0.005	0.01	0.015	0.02	0.025	0.05	0.10	0.20
29	11.0	13.1	14.3	15.0	15.6	16.0	17.7	19.8	22.5
30	11.6	13.8	15.0	15.7	16.3	16.8	18.5	20.6	23.4
32	12.8	15.1	16.4	17.2	17.8	18.3	20.1	22.3	25.1
34	14.1	16.5	17.8	18.6	19.3	19.8	21.7	24.0	26.9
36	15.3	17.9	19.2	20.1	20.8	21.3	23.3	25.6	28.7
38	16.6	19.3	20.7	21.6	22.3	22.9	24.9	27.3	30.5
40	17.9	20.7	22.2	23.1	23.8	24.4	26.5	29.1	32.3
42	19.2	22.1	23.7	24.6	25.4	26.0	28.1	30.8	34.2
44	20.6	23.6	25.1	26.2	26.9	27.6	29.8	32.5	36.0
46	21.9	25.0	26.7	27.7	28.5	29.2	31.4	34.2	37.8
48	23.3	26.5	28.2	29.3	30.1	30.8	33.1	35.9	39.6
50	24.7	28.0	29.7	30.8	31.7	32.4	34.8	37.7	41.5
55	28.2	31.7	33.6	34.8	35.7	36.4	39.0	42.1	46.0
60	31.7	35.5	37.5	38.7	39.7	40.5	43.2	46.5	50.6
65	35.4	39.4	41.4	42.8	43.8	44.6	47.4	50.9	55.3
70	39.0	43.3	45.4	46.8	47.9	48.8	51.7	55.3	59.9
75	42.8	47.2	49.5	50.9	52.0	52.9	56.1	59.8	64.5
80	46.5	51.2	53.5	55.1	56.2	57.1	60.4	64.3	69.2
85	50.3	55.2	57.6	59.2	60.4	61.4	64.7	68.8	73.9
90	54.2	59.2	61.7	63.4	64.6	65.6	69.1	73.3	78.6
95	58.0	63.2	65.9	67.6	68.9	69.9	73.5	77.8	83.2
100	61.9	67.3	70.1	71.8	73.1	74.2	77.9	82.4	87.9

(b) *Upper percentiles.* This table gives values $x(\alpha)$ defined by the equation

$$\text{Prob}\{\chi_v^2 \geq x(\alpha)\} = \alpha$$

for selected values of the degrees of freedom v and selected values of α as shown in Figure A3(b). For example, the upper 5% point of the χ_5^2 distribution is 11.1, corresponding to $1 - \alpha = 0.95$.

					$1 - \alpha$				
$x(\alpha)$	0.80	0.90	0.95	0.975	0.98	0.985	0.99	0.995	0.999
$v = 1$	1.64	2.71	3.84	5.03	5.41	5.92	6.64	7.88	10.8
2	3.22	4.61	5.99	7.38	7.82	8.40	9.21	10.6	13.8
3	4.64	6.25	7.81	9.35	9.84	10.5	11.3	12.8	16.3
4	5.99	7.78	9.49	11.1	11.7	12.3	13.3	14.9	18.5
5	7.29	9.24	11.1	12.8	13.4	14.1	15.1	16.7	20.5
6	8.56	10.6	12.6	14.4	15.0	15.8	16.8	18.5	22.5
7	9.80	12.0	14.1	16.0	16.6	17.4	18.5	20.3	24.3
8	11.0	13.4	15.5	17.5	18.2	19.0	20.1	22.0	26.1

$$1 - \alpha$$

$x(\alpha)$	0.80	0.90	0.95	0.975	0.98	0.985	0.99	0.995	0.999
9	12.2	14.7	16.9	19.0	19.7	20.5	21.7	23.6	27.9
10	13.4	16.0	18.3	20.5	21.2	22.0	23.2	25.2	29.6
11	14.6	17.3	19.7	21.9	22.6	23.5	24.7	26.8	31.3
12	15.8	18.5	21.0	23.3	24.1	25.0	26.2	28.3	32.9
13	17.0	19.8	22.4	24.7	25.5	26.4	27.7	29.8	34.5
14	18.2	21.1	23.7	26.1	26.9	27.8	29.1	31.3	36.1
15	19.3	22.3	25.0	27.5	28.3	29.2	30.6	32.8	37.7
16	20.5	23.5	26.3	28.8	29.6	30.6	32.0	34.3	39.3
17	21.6	24.8	27.6	30.2	31.0	32.0	33.4	35.7	40.8
18	22.8	26.0	28.9	31.5	32.4	33.4	34.8	37.2	42.3
19	23.9	27.2	30.1	32.9	33.7	34.7	36.2	38.6	43.8
20	25.0	28.4	31.4	34.2	35.0	36.1	37.6	40.0	45.3
21	26.2	29.6	32.7	35.5	36.3	37.4	38.9	41.4	46.8
22	27.3	30.8	33.9	36.8	37.7	38.8	40.3	42.8	48.3
23	28.4	32.0	35.2	38.1	39.0	40.1	41.6	44.2	49.7
24	29.6	33.2	36.4	39.4	40.3	41.4	43.0	45.6	51.2
25	30.7	34.4	37.7	40.7	41.6	42.7	44.3	46.9	52.6
26	31.8	35.6	38.9	41.9	42.9	44.0	45.6	48.3	54.1
27	32.9	36.7	40.1	43.2	44.1	45.3	47.0	49.6	55.5
28	34.0	37.9	41.3	44.5	45.4	46.6	48.3	51.0	56.9
29	35.1	39.1	42.6	45.7	46.7	47.9	49.6	52.3	58.3
30	36.2	40.3	43.8	47.0	48.0	49.2	50.9	53.7	59.7
32	38.5	42.6	46.2	49.5	50.5	51.8	53.5	56.3	62.5
34	40.7	44.9	48.6	52.0	53.0	54.3	56.1	59.0	65.2
36	42.9	47.2	51.0	54.4	55.5	56.8	58.6	61.6	68.0
38	45.1	49.5	53.4	56.9	58.0	59.3	61.2	64.2	70.7
40	47.3	51.8	55.8	59.3	60.4	61.8	63.7	66.8	73.4
42	49.5	54.1	58.1	61.8	62.9	64.3	66.2	69.3	76.1
44	51.6	56.4	60.5	64.2	65.3	66.8	68.7	71.9	78.7
46	53.8	58.6	62.8	66.6	67.8	69.2	71.2	74.4	81.4
48	56.0	60.9	65.2	69.0	70.2	71.7	73.7	77.0	84.0
50	58.2	63.2	67.5	71.4	72.6	74.1	76.2	79.5	86.7
55	63.6	68.8	73.3	77.4	78.6	80.2	82.3	85.7	93.2
60	69.0	74.4	79.1	83.3	84.6	86.2	88.4	92.0	99.6
65	74.3	80.0	84.8	89.2	90.5	92.2	94.4	98.1	106
70	79.7	85.5	90.5	95.0	96.4	98.1	100	104	112
75	85.1	91.1	96.2	101	102	104	106	110	119
80	90.4	96.6	102	107	108	110	112	116	125
85	95.7	102	108	112	114	116	118	122	131
90	101	108	113	118	120	122	124	128	137
95	106	113	119	124	125	127	130	134	143
100	112	119	124	130	131	133	136	140	149

A.3 Values of $y = A_1^{-1}(x)$ for $0 \le x \le 1$

This table gives solutions of the equation $A_1(y) = x$, where the function $A_1(y) \equiv I_1(y)/I_0(y)$ is the ratio of two modified Bessel functions, as defined in §**3.3.6** *Note 1(i)*.

If x is the mean resultant length ρ of a von Mises distribution, then y is the value of the concentration parameter κ of this distribution.

If x is the mean resultant length $\hat{\rho} = \bar{R}$ of a random sample from a von Mises distribution with concentration parameter κ, then y is the corresponding maximum likelihood estimate $\hat{\kappa}_{ML}$ of κ.

x	$A_1^{-1}(x)$	x	$A_1^{-1}(x)$	x	$A_1^{-1}(x)$	x	$A_1^{-1}(x)$	x	$A_1^{-1}(x)$
0.0005	0.001	0.22	0.451	0.47	1.07	0.72	2.14	0.930	7.43
		0.23	0.473	0.48	1.10	0.73	2.21	0.935	7.97
0.001	0.002	0.24	0.495	0.49	1.13	0.74	2.29	0.940	8.61
0.005	0.010	0.25	0.516	0.50	1.16	0.75	2.37	0.945	9.37
0.01	0.020	0.26	0.539	0.51	1.19	0.76	2.46	0.950	10.3
0.02	0.040	0.27	0.561	0.52	1.22	0.77	2.55	0.955	11.4
0.03	0.060	0.28	0.584	0.53	1.26	0.78	2.65	0.960	12.8
0.04	0.080	0.29	0.606	0.54	1.29	0.79	2.76	0.965	14.6
0.05	0.100	0.30	0.629	0.55	1.33	0.80	2.87	0.970	16.9
0.06	0.120	0.31	0.652	0.56	1.36	0.81	3.00	0.975	20.3
0.07	0.140	0.32	0.676	0.57	1.40	0.82	3.15	0.980	25.3
0.08	0.161	0.33	0.700	0.58	1.44	0.83	3.32	0.985	33.6
0.09	0.181	0.34	0.724	0.59	1.48	0.84	3.51	0.990	50.3
0.10	0.201	0.35	0.748	0.60	1.52	0.85	3.74		
0.11	0.221	0.36	0.772	0.61	1.56	0.86	3.91	0.991	55.8
0.12	0.242	0.37	0.797	0.62	1.60	0.87	4.18	0.992	62.7
0.13	0.262	0.38	0.823	0.63	1.65	0.88	4.49	0.993	71.7
0.14	0.283	0.39	0.848	0.64	1.69	0.89	4.86	0.994	83.6
0.15	0.303	0.40	0.874	0.65	1.74	0.90	5.30	0.995	100
0.16	0.324	0.41	0.900	0.66	1.79			0.996	125
0.17	0.345	0.42	0.927	0.67	1.84	0.905	5.56	0.997	167
0.18	0.366	0.43	0.954	0.68	1.90	0.910	5.85	0.998	250
0.19	0.387	0.44	0.982	0.69	1.95	0.915	6.18	0.999	500
0.20	0.408	0.45	1.01	0.70	2.01	0.920	6.54		
0.21	0.430	0.46	1.04	0.71	2.08	0.925	6.95	0.9995	1000

A.4 Values of $x = A_1(y)$ for $y \geq 0$

This table gives values of the ratio $A_1(y) = I_1(y)/I_0(y)$, where $I_0(y)$ and $I_1(y)$ are two modified Bessel functions, as defined in §**3.3.6** *Note 1(i)*.

If y is the concentration parameter (κ) of a von Mises distribution, then x is the value of the mean resultant length (ρ) of this distribution.

If y is the maximum likelihood estimate ($\widehat{\kappa}_{ML}$) of the concentration parameter κ, based on a random sample from a von Mises distribution, then x is the value of the sample mean resultant length $\widehat{\rho} = \bar{R}$.

y	$A_1(y)$	y	$A_1(y)$	y	$A_1(y)$	y	$A_1(y)$	y	$A_1(y)$
0.001	0.0005	0.30	0.1483	1.15	0.4970	2.0	0.6978	6.0	0.9124
0.005	0.0025	0.35	0.1724	1.20	0.5128			7.0	0.9255
0.01	0.0050	0.40	0.1961	1.25	0.5280	2.1	0.7135	8.0	0.9352
		0.45	0.2195	1.30	0.5427	2.2	0.7280	9.0	0.9427
0.02	0.0100	0.50	0.2425	1.35	0.5568	2.3	0.7414	10.0	0.9486
0.03	0.0150	0.55	0.2651	1.40	0.5704	2.4	0.7536		
0.04	0.0200	0.60	0.2873	1.45	0.5835	2.5	0.7649	15.0	0.9661
0.05	0.0250	0.65	0.3090	1.50	0.5961	2.6	0.7754	20.0	0.9747
0.06	0.0300	0.70	0.3302	1.55	0.6083	2.7	0.7850		
0.07	0.0350	0.75	0.3509	1.60	0.6199	2.8	0.7939	30.0	0.9832
0.08	0.0400	0.80	0.3711	1.65	0.6311	2.9	0.8021	40.0	0.9874
0.09	0.0450	0.85	0.3907	1.70	0.6418	3.0	0.8096	50.0	0.9899
0.10	0.0499	0.90	0.4098	1.75	0.6521				
		0.95	0.4284	1.80	0.6620	3.5	0.8397	100.0	0.9950
0.15	0.0748	1.00	0.4464	1.85	0.6715	4.0	0.8635		
0.20	0.0995	1.05	0.4638	1.90	0.6806	4.5	0.8803	500.0	0.9990
0.25	0.1240	1.10	0.4807	1.95	0.6894	5.0	0.8934		

A.5 Selected percentiles for the statistic V

The tabulated values are upper percentage points of the modified Kuiper statistic V.

	Percentage point $100\alpha\%$			
α	0.15	0.10	0.05	0.01
V	1.537	1.620	1.747	2.001

Source: Stephens (1974, Table 1A).

A.6 Selected confidence intervals for the median direction

Sample size	Interval	Confidence level (%)
3	$(\theta_{(1)}, \theta_{(3)})$	75.0
4	$(\theta_{(1)}, \theta_{(4)})$	87.5
5	$(\theta_{(1)}, \theta_{(5)})$	93.7
6	$(\theta_{(1)}, \theta_{(6)})$	97.0
7	$(\theta_{(1)}, \theta_{(7)})$	98.4
	$(\theta_{(2)}, \theta_{(6)})$	87.5
8	$(\theta_{(1)}, \theta_{(8)})$	99.2
	$(\theta_{(2)}, \theta_{(7)})$	93.0
9	$(\theta_{(1)}, \theta_{(9)})$	99.6
	$(\theta_{(2)}, \theta_{(8)})$	96.1
10	$(\theta_{(2)}, \theta_{(9)})$	97.8
	$(\theta_{(3)}, \theta_{(8)})$	89.3
11	$(\theta_{(2)}, \theta_{(10)})$	99.0
	$(\theta_{(3)}, \theta_{(9)})$	93.4
12	$(\theta_{(3)}, \theta_{(10)})$	96.2
	$(\theta_{(4)}, \theta_{(9)})$	85.4
13	$(\theta_{(3)}, \theta_{(11)})$	97.8
	$(\theta_{(4)}, \theta_{(10)})$	92.8
14	$(\theta_{(4)}, \theta_{(11)})$	93.7
15	$(\theta_{(3)}, \theta_{(13)})$	96.5

Source: Fisher & Powell (1989, Table 9).

A.7 Critical values for the Wilcoxon signed-rank statistic

For a given sample size n, the table gives the probability Prob $(w_n^+ \geq x)$ for an observed value x of the Wilcoxon signed-rank statistic.

				n				
x	3	4	5	6	7	8	9
3	.625						
4	.375						
5	.250	.562					
6	.125	.438					
7		.312					
8		.188	.500				
9		.125	.406				
10		.062	.312				
11			.219	.500			
12			.156	.422			
13			.094	.344			
14			.062	.281	.531		
15			.031	.219	.469		
16				.156	.406		
17				.109	.344		
18				.078	.289	.527	
19				.047	.234	.473	
20				.031	.188	.422	
21				.016	.148	.371	
22					.109	.320	
23					.078	.273	.500
24					.055	.230	.455
25					.039	.191	.410
26					.023	.156	.367
27					.016	.125	.326
28					.008	.098	.285
29						.074	.248
30						.055	.213
31						.039	.180
32						.027	.150
33						.020	.125
34						.012	.102
35						.008	.082
36						.004	.064
37							.049
38							.037
39							.027
40							.020
41							.014
42							.010
43							.006
44							.004
45							.002

x	10	11	12	13	14	15
			n			
28	.500					
29	.461					
30	.423					
31	.385					
32	.348					
33	.312	.517				
34	.278	.483				
35	.246	.449				
36	.216	.416				
37	.188	.382				
38	.161	.350				
39	.138	.319	.515			
40	.116	.289	.485			
41	.097	.260	.455			
42	.080	.232	.425			
43	.065	.207	.396			
44	.053	.183	.367			
45	.042	.160	.339			
46	.032	.139	.311	.500		
47	.024	.120	.285	.473		
48	.019	.103	.259	.446		
49	.014	.087	.235	.420		
50	.010	.074	.212	.393		
51	.007	.062	.190	.368		
52	.005	.051	.170	.342		
53	.003	.042	.151	.318	.500	
54	.002	.034	.133	.294	.476	
55	.001	.027	.117	.271	.452	
56		.021	.102	.249	.428	
57		.016	.088	.227	.404	
58		.012	.076	.207	.380	
59		.009	.065	.188	.357	
60		.007	.055	.170	.335	.511
61		.005	.046	.153	.313	.489
62		.003	.039	.137	.292	.467
63		.002	.032	.122	.271	.445
64		.001	.026	.108	.251	.423
65		.001	.021	.095	.232	.402
66		.000	.017	.084	.213	.381
67			.013	.073	.196	.360
68			.010	.064	.179	.339
69			.008	.055	.163	.319
70			.006	.047	.148	.300
71			.005	.040	.134	.281
72			.003	.034	.121	.262
73			.002	.029	.108	.244
74			.002	.024	.097	.227
75			.001	.020	.086	.211
76			.001	.016	.077	.195
77			.000	.013	.068	.180

x	n					
	10	11	12	13	14	15
78			.000	.011	.059	.165
79				.009	.052	.151
80				.007	.045	.138
81				.005	.039	.126
82				.004	.034	.115
83				.003	.029	.104
84				.002	.025	.094
85				.002	.021	.084
86				.001	.018	.076
87				.001	.015	.068
88				.001	.012	.060
89				.000	.010	.053
90				.000	.008	.047
91				.000	.007	.042
92					.005	.036
93					.004	.032
94					.003	.028
95					.003	.024
96					.002	.021
97					.002	.018
98					.001	.015
99					.001	.013
100					.001	.011
101					.000	.009
102					.000	.008
103					.000	.006
104					.000	.005
105					.000	.004
106						.003
107						.003
108						.002
109						.002
110						.001
111						.001
112						.001
113						.001
114						.000
115						.000
116						.000
117						.000
118						.000
119						.000
120						.000

Source: Hollander & Wolfe (1973, Table A.4), reproduced with kind permission of the authors and John Wiley & Sons.

A.8 Critical values for testing goodness of fit of a von Mises distribution

The values in the body of the table are selected upper $100\alpha\%$ points of the distribution of the statistic U^2 for testing goodness of fit of a von Mises $VM(\mu, \kappa)$ distribution to a random sample of n data values. Different parts of the table should be used, depending on whether μ or κ is known.

κ	Significance level α							
	0·500	0·25	0·15	0·10	0·05	0·025	0·01	0·005
	Case 0: both parameters known							
All κ	0·069	0·105	0·131	0·152	0·187	0·222	0·268	0·304
	Case 1: μ known, κ unknown							
0·0	0·047	0·071	0·089	0·105	0·133	0·163	0·204	0·235
0·50	0·048	0·072	0·091	0·107	0·135	0·165	0·205	0·237
1·00	0·051	0·076	0·095	0·111	0·139	0·169	0·209	0·241
1·50	0·053	0·080	0·099	0·115	0·144	0·173	0·214	0·245
2·00	0·055	0·082	0·102	0·119	0·147	0·177	0·217	0·248
4·00	0·058	0·086	0·107	0·124	0·153	0·183	0·224	0·255
∞	0·059	0·089	0·110	0·127	0·157	0·187	0·228	0·259
	Case 2: μ unknown, κ known							
0·0	0·047	0·071	0·089	0·105	0·133	0·163	0·204	0·235
0·50	0·048	0·072	0·091	0·107	0·135	0·165	0·205	0·237
1·00	0·051	0·076	0·095	0·111	0·139	0·169	0·209	0·241
1·50	0·053	0·080	0·100	0·116	0·144	0·174	0·214	0·245
2·00	0·055	0·082	0·103	0·119	0·148	0·177	0·218	0·249
4·00	0·057	0·085	0·106	0·122	0·151	0·181	0·221	0·253
∞	0·057	0·085	0·105	0·122	0·151	0·180	0·221	0·252
	Case 3: μ unknown, κ unknown							
0·0	0·030	0·040	0·046	0·052	0·061	0·069	0·081	0·090
0·50	0·031	0·042	0·050	0·056	0·066	0·077	0·090	0·100
1·00	0·035	0·049	0·059	0·066	0·079	0·092	0·110	0·122
1·50	0·039	0·056	0·067	0·077	0·092	0·108	0·128	0·144
2·00	0·043	0·061	0·074	0·084	0·101	0·119	0·142	0·159
4·00	0·047	0·067	0·082	0·093	0·113	0·132	0·158	0·178
∞	0·048	0·069	0·084	0·096	0·117	0·137	0·164	0·184

For $\kappa > 4$ use linear interpolation in $1/\kappa$. For Cases 2 and 3 enter the table at the estimate of κ.

Source: Lockhart & Stephens (1985, Table 1), reproduced with kind permission of the authors and the *Biometrika Trust*.

A.9 Critical values for the discordancy test for a von Mises sample

The values in the body of the table are the upper 5% and 1% critical values of the statistic M_n, for testing a single outlier in a von Mises random sample of size n for discordancy.

						Sample size n					
α	3	4	5	6	7	8	9	10	12	15	20
0.05	0.661	0.73	0.731	0.714	0.680	0.648	0.611	0.583	0.528	0.464	0.387
0.01	0.661	0.75	0.774	0.776	0.763	0.743	0.708	0.683	0.629	0.564	0.474

A.10 Critical values for the angular–linear rank correlation statistic U_n

The values in the body of the table are the upper $100\alpha\%$ critical values of the statistic U_n, for testing C-association between a linear random variable X and a circular random variable Θ, for a random sample of size n.

	α		
n	0.10	0.05	0.01
5	3.97	–	–
6	4.57	4.67	4.95
7	4.30	4.90	5.75
8	4.488	5.17	6.15
9	4.50	5.34	6.64
10	4.52	5.48	6.68
11	4.55	5.5	7.2
12	4.57	5.6	7.5
15	4.59	5.7	7.9
20	4.60	5.8	8.3
30	4.60	5.9	8.7
40	4.60	5.9	8.8
50	4.61	6.0	8.9
100	4.61	6.0	9.1

Source: Adapted from Mardia (1976, Table 1).

A.11 Assessment of significance of the angular–linear rank correlation coefficients $\widehat{\lambda}_n$ and $\widehat{\lambda}_{n,M}$

These tables provide distributions or critical values, depending on sample size, of the rank correlation statistic $\widehat{\lambda}_n$, for testing C-association between a linear random variable X and a circular random variable Θ, for a random sample of size n.

(a) $n < 9$. The table entries give selected cumulative probabilities $\text{Prob}(\widehat{\lambda}_n \le x)$ for $n = 6, 7, 8$.

$n = 6$		$n = 7$		$n = 8$	
x	$\text{Prob}(\widehat{\lambda}_n \le x)$	x	$\text{Prob}(\widehat{\lambda}_n \le x)$	x	$\text{Prob}(\widehat{\lambda}_n \le x)$
0.67	0.69	0.80	0.87	0.80	0.90
0.73	0.83	0.83	0.91	0.87	0.96
0.80	0.87	0.88	0.96	0.93	0.98

(b) $n \ge 9$: critical values for incomplete statistic $\widehat{\lambda}_{n,M}$. To compare the incomplete statistic $\widehat{\lambda}_{n,M}$ (defined by equation (6.18)) with the upper $100\alpha\%$ point of its null distribution, calculate $\eta = n^2/M$ and

$$L(\alpha) = n(\widehat{\lambda}_{n,M} - 2/3) - K_1(\alpha) - \eta/3 - (K_2(\alpha) + \eta K_3(\alpha))/n$$

where the coefficients $K_1(\alpha), K_2(\alpha)$ and $K_3(\alpha)$ are given in the table below. The hypothesis of no C-association between X and Θ is rejected if $L(\alpha) > 0$.

	Coefficients		
α	$K_1(\alpha)$	$K_2(\alpha)$	$K_3(\alpha)$
0.10	0.518	4.787	−9.231
0.05	0.774	6.174	−9.884
0.01	1.369	8.012	−6.215

(c) $n \geq 20$: asymptotic distribution. The table gives selected upper $100\alpha\%$ critical values for the statistic $\widehat{\Lambda} = n(\widehat{\lambda}_n - 2/3)$. The hypothesis of no C-association between X and Θ is rejected if $\widehat{\Lambda}$ is too large.

	Percentage point ($100\alpha\%$)					
α	0.20	0.10	0.05	0.25	0.01	0.001
$\widehat{\Lambda}$	0.261	0.518	0.774	1.031	1.369	2.221

Source: Fisher & Lee (1981).

A.12 Critical values for the angular–angular rank correlation coefficient $\widehat{\Delta}_n$

These tables provide distributions or critical values, depending on sample size, of the rank correlation statistic $\widehat{\Delta}_n$, for testing C-linear association between a circular random variable Θ and another circular random variable Φ, for a random sample of size n.

(a) $n < 8$. The probability distribution of $n\widehat{\Delta}_n$ is given in the following table, for $n = 3,\ldots,7$. Note that the distribution is symmetric about 0, under the hypothesis that $\Delta = 0$.

$n=3$											
x	3	-3									
$2\,\mathrm{Prob}(3\widehat{\Delta}_3 = x)$	1	1									
$n=4$											
x	4	0	-4								
$6\,\mathrm{Prob}(4\widehat{\Delta}_4 = x)$	1	4	1								
$n=5$											
x	5	2	1	0	$\ldots$						
$24\,\mathrm{Prob}(5\widehat{\Delta}_5 = x)$	1	5	5	2							
$n=6$											
x	6.0	3.6	2.4	1.2	0.0	$\ldots$					
$120\,\mathrm{Prob}(6\widehat{\Delta}_6 = x)$	1	6	12	23	36						
$n=7$											
x	7.0	5.0	3.8	3.4	2.6	2.2	1.8	1.4	1.0	0.6	0.2
$720\,\mathrm{Prob}(7\widehat{\Delta}_7 = x)$	1	7	14	21	14	21	63	44	28	70	77

(b) $n \geq 9$. The values in the body of the table are selected upper $100\alpha\%$ points of distribution of the statistic $n\widehat{\Delta}_n$.

	α					
n	0.10	0.05	0.025	0.01	0.005	0.001
8	2.10	2.78	3.40	4.25	4.74	6.23
9	2.07	2.72	3.33	4.16	4.64	6.08
10	2.04	2.68	3.28	4.09	4.56	5.96
11	2.01	2.65	3.24	4.02	4.50	5.85
12	1.99	2.62	3.20	3.97	4.44	5.77
13	1.97	2.59	3.17	3.93	4.40	5.70
14	1.96	2.57	3.14	3.89	4.36	5.64
15	1.95	2.56	3.12	3.86	4.33	5.59
20	1.90	2.49	3.04	3.75	4.21	5.40
25	1.87	2.46	2.99	3.68	4.14	5.29
30	1.86	2.43	2.96	3.64	4.09	5.22
∞	1.77	2.31	2.81	3.42	3.85	4.85

Source: Fisher & Lee (1982).

A.13 Critical values for the angular–angular rank correlation coefficient $\widehat{\Pi}_n$

These tables provide distributions or critical values, depending on sample size, of the rank correlation statistic $\widehat{\Pi}_n$, for testing C-linear association between a circular random variable Θ and another circular random variable Φ, for a random sample of size n.

(a) $n < 8$. The probability distribution of $(n-1)\widehat{\Pi}_n$ is given in the following table, for $n = 3, \ldots, 7$. Note that the distribution is symmetric about 0, under the hypothesis of no C-linear association.

$n = 3$

x	2	-2
$2\,\mathrm{Prob}(3\widehat{\Pi}_3 = x)$	1	1

$n = 4$

x	3	0	-3
$6\,\mathrm{Prob}(3\widehat{\Pi}_4 = x)$	1	4	1

$n = 5$

x	4	1.79	0	$\cdots$
$24\,\mathrm{Prob}(4\widehat{\Pi}_5 = x)$	1	5	12	

$n = 6$

x	60	40	25	20	15	5	0	$\cdots$
$120\,\mathrm{Prob}(5\widehat{\Pi}_6 = x)$	1	6	6	3	14	12	36	

$n = 7$

x	6	4.71	3.68	3.47	2.78	2.71	1.93	1.83	1.81	1.54	1.33
$720\,\mathrm{Prob}(6\widehat{\Pi}_7 = x)$	1	7	7	7	14	7	28	7	7	28	14

x	1.07	0.86	0.69	0.59	0.52	0.48	0.38	0.31	0.71	0	$\cdots$
$720\,\mathrm{Prob}(6\widehat{\Pi}_7 = x)$	28	42	28	7	7	28	28	7	14	88	

(b) $n \geq 9$. The values in the body of the table are selected upper $100\alpha\%$ points of distribution of the statistic $(n-1)\widehat{\Pi}_n$.

			α		
n	0.10	0.05	0.025	0.01	0.005
8	1.80	2.55	3.26	4.11	4.80
9	1.78	2.52	3.23	4.09	4.78
10	1.76	2.50	3.21	4.07	4.77
11	1.74	2.48	3.19	4.06	4.75
12	1.73	2.46	3.17	4.04	4.74
13	1.72	2.45	3.15	4.03	4.73
14	1.71	2.44	3.14	4.02	4.72
15	1.71	2.43	3.13	4.02	4.71
20	1.68	2.40	3.10	3.99	4.67
25	1.67	2.38	3.07	3.97	4.64
30	1.66	2.36	3.06	3.96	4.63
∞	1.61	2.30	2.99	3.91	4.60

Source: Fisher & Lee (1983).

Appendix B: Data sets

B.1 Arrival times at an intensive care unit

Data Arrival times on a 24-hour clock of 254 patients at an intensive care unit, over a period of about 12 months.

Type Vectors.

Source Cox & Lewis (1966, pp. 254–5).

Analysis Examples 2.1, 2.3, 2.4, 2.5, 2.6, 2.8, 2.10.

11.00	17.00	23.15	10.00	12.00	08.45	16.00	10.00	15.30	20.20
04.00	12.00	02.20	12.00	05.30	07.30	12.00	16.00	16.00	01.30
11.05	16.00	19.00	17.45	20.20	21.00	12.00	12.00	18.00	22.00
22.00	22.05	12.45	19.30	18.45	16.15	16.00	20.30	23.40	20.20
18.45	16.30	22.00	08.45	19.15	15.30	12.00	18.15	14.00	13.00
23.00	19.15	22.00	10.15	12.30	18.15	21.05	21.00	00.30	01.45
12.20	14.45	22.30	12.30	13.15	17.30	11.20	17.30	23.00	10.55
13.30	11.00	18.30	11.05	04.00	07.30	20.00	21.30	06.30	17.30
20.45	22.00	20.15	21.00	17.30	19.50	02.00	01.45	03.40	04.15
23.55	03.15	19.00	21.45	21.30	00.45	02.30	15.30	21.00	08.45
14.30	17.00	03.30	15.45	17.30	14.00	02.00	11.30	17.30	17.10
21.20	03.00	13.30	23.00	20.10	23.15	20.00	16.00	18.30	21.00
21.10	17.00	13.25	15.05	14.10	19.15	14.05	22.40	09.30	17.30
12.30	17.30	14.30	16.00	14.10	14.00	15.30	04.30	11.50	11.55
15.20	15.40	11.15	02.15	11.15	21.30	03.00	00.40	10.00	09.45
23.45	10.00	07.50	13.30	12.30	13.45	19.30	00.15	07.45	15.20
18.40	19.50	23.55	01.45	10.50	07.50	15.30	18.00	23.05	19.30
19.00	16.10	10.00	02.30	22.00	21.50	19.10	11.45	15.45	16.30
18.30	10.05	20.00	13.35	16.45	02.15	20.30	14.00	21.15	18.45
14.05	14.15	01.15	01.45	18.00	14.15	15.15	16.15	10.20	13.35
17.15	19.50	22.45	07.25	17.00	12.30	23.15	10.30	13.45	02.30
12.00	15.45	17.00	17.00	01.30	20.15	12.30	15.40	03.30	18.35
13.30	16.40	18.00	20.00	11.15	16.40	13.55	21.00	07.45	22.30
16.40	23.10	19.15	11.00	00.15	14.40	15.45	12.45	17.00	18.00
21.45	16.00	12.00	02.30	12.55	20.20	10.30	15.50	17.30	20.00
02.00	01.45	01.45	02.05						

B.2 Long-axis orientations of feldspar laths

Data   Measurements of long-axis orientation of 133 feldspar laths in basalt.

Type   Axes.

Source   Smith (1988, set 28-6-1co.prn); data kindly supplied by Ms Nicola Smith.

Analysis   Examples 2.2, 2.7, 2.8.

176	162	49	174	174	49	54	63	59	61
66	104	97	58	121	5	178	3	168	0
18	39	140	63	55	170	169	37	152	73
53	176	72	170	113	56	87	161	164	21
50	6	59	140	54	64	56	38	61	143
51	144	148	44	60	98	86	145	38	168
39	134	68	57	129	68	132	82	54	119
131	50	93	160	127	124	65	108	52	61
86	37	132	83	163	58	144	29	80	172
144	138	10	45	137	11	145	103	69	124
54	121	1	39	111	153	13	5	5	107
104	39	133	36	63	4	21	51	30	52
90	143	13	50	109	12	170	5	14	91
132	12	1							

B.3 Movements of turtles

Data   Measurements of the directions taken by 76 turtles after treatment.

Type   Vectors.

Source   Stephens (1969b).

Analysis   Examples 2.9, 4.25.

8	9	13	13	14	18	22	27	30	34
38	38	40	44	45	47	48	48	48	48
50	53	56	57	58	58	61	63	64	64
64	65	65	68	70	73	78	78	78	83
83	88	88	88	90	92	92	93	95	96
98	100	103	106	113	118	138	153	153	155
204	215	223	226	237	238	243	244	250	251
257	268	285	319	343	350				

B.4 Directional preferences of starhead topminnows

Data   Sun compass orientations of 50 starhead topminnows, measured under heavily overcast conditions.

Type   Vectors.

Source   Goodyear (1970, Figure 1D).

Analysis   Examples 4.1, 4.12.

2	9	18	24	30	35	35	39	39	44
44	49	56	70	76	76	81	86	91	112
121	127	133	134	138	147	152	157	166	171
177	187	206	210	211	215	238	246	269	270
285	292	305	315	325	328	329	343	354	359

B.5 Long-axis orientations of feldspar laths

Data   Measurements of long-axis orientation of 164 feldspar laths in basalt.

Type   Axes.

Source   Randomly selected from Smith (1988, set 24-6-5co.prn); data kindly supplied by Ms Nicola Smith.

Analysis   Examples 4.2, 4.5, 4.7, 4.8, 4.11.

1	1	2	2	3	8	9	12	16	17
19	23	28	28	34	34	35	36	36	37
41	45	49	50	51	53	58	68	69	70
72	72	76	78	80	85	97	97	99	101
105	121	125	126	133	141	143	149	152	156
160	163	167	168	170	171	172	174	175	176

B.6 Cross-bed azimuths of palaeocurrents

Data   Sets of cross-bed azimuths of palaeocurrents measured in the Belford Anticline (New South Wales).

Type   Vectors.

Source   Fisher & Powell (1989).

Analysis   Examples 4.3, 4.6, 4.13, 4.14, 4.16, 4.17, 5.1, 5.11.

Set 1 : n = 40

284	311	334	320	294	137	123	166	143	127
244	243	152	242	143	186	263	234	209	267
315	329	235	38	241	319	308	127	217	245
169	161	263	209	228	168	98	278	154	279

Set 2 : n = 30

294	301	329	315	277	281	254	245	272	242
177	257	177	229	250	166	232	245	224	186
257	267	241	239	287	229	290	214	215	224

Set 3 : $n = 30$

163	275	218	287	313	322	236	254	239	286
268	245	211	271	151	309	27	224	181	220
217	192	292	283	216	231	147	163	155	203

B.7 Movements of ants

Data Directions chosen by 100 ants in response to an evenly illuminated black target placed as shown.

Type Vectors.

Source Randomly selected values from Jander (1957, Figure 18A).

Analysis Examples 4.4, 4.18, 4.19.

330	290	60	200	200	180	280	220	190	180
180	160	280	180	170	190	180	140	150	150
160	200	190	250	180	30	200	180	200	350
200	180	120	200	210	130	30	210	200	230
180	160	210	190	180	230	50	150	210	180
190	210	220	200	60	260	110	180	220	170
10	220	180	210	170	90	160	180	170	200
160	180	120	150	300	190	220	160	70	190
110	270	180	200	180	140	360	150	160	170
140	40	300	80	210	200	170	200	210	190

B.8 Orientations of pebbles

Data Horizontal axes of 100 outwash pebbles from a late Wisconsin outwash terrace along Fox River, near Cary, Illinois.

Type Axes.

Source Mardia (1972a, Table 1.6), adapted from Krumbein (1939).

Analysis Example 4.9.

Direction	0	20	40	60	80	100	120	140	160
Frequency	16	13	9	14	9	14	12	6	7

B.9 Dance directions of bees

Data   Dance directions of 279 honey bees viewing a zenith patch
            of artificially polarised light.

Type   Vectors.

Source   Adapted from Wehner & Strasser (1985, p. 346),
            based on advice from Prof. Dr Wehner.

Analysis Example 4.10.

Direction	0	10	20	30	40	50	60	70	80	90
Frequency	3	8	9	9	6	6	12	9	9	9

Direction	100	110	120	130	140	150	160	170	180	190
Frequency	9	12	5	6	8	12	8	9	12	5

Direction	200	210	220	230	240	250	260	270	280	290
Frequency	5	9	8	5	12	9	8	7	3	8

Direction	300	310	320	330	340	350
Frequency	12	6	5	5	8	3

B.10 Directions of desert ants

Data   Directions of 11 long-legged desert ants (*Cataglyphis fortis*)
            after one eye on each ant was 'trained' to learn the ant's home
            direction, then covered and the other eye uncovered.

Type   Vectors.

Source   Prof. Dr R. Wehner (Personal communication);
            experiment described in Wehner & Müller (1985).

Analysis Examples 4.15, 5.10.

Set 1 : n = 11

11	3	−22	−1	−7	27	−2	15	14	−13
0									

Set 2 : n = 32

−46	10	−3	−9	19	−5	49	24	14	14
4	−14	−4	30	4	−172	24	−6	−32	−128
−68	−12	8	21	10	−11	−12	25	24	−7
18	−3								

Set 3 : n = 18

22	32	−25	4	43	108	47	−49	−67	−19
−14	4	−2	140	82	6	−21	19		

B.11 Movements of sea stars

Data Resultant directions of 22 sea stars 11 days after being displaced from their natural habitat.

Type Vectors.

Source Upton & Fingleton (1989, p. 274), as adapted from Pabst & Vicentini (1978).

Analysis Example 4.20.

0	1	3	3	8	13	16	18	30	31
43	45	147	298	329	332	335	340	350	354
356	357								

B.12 Vanishing directions of homing pigeons

Data Vanishing directions of 15 homing pigeons, released just over 16 kilometres Northwest of their loft.

Type Vectors.

Source Schmidt-Koenig (1963).

Analysis Examples 4.21, 4.22, 4.23, 4.24.

85	135	135	140	145	150	150	150	160	285
200	210	220	225	270					

B.13 Orientations of termite mounds

Data   Orientations of termite mounds of *Amitermes laurensis* at 14 sites in Cape York Peninsula, North Queensland.

Type   Axes.

Source   Data kindly supplied by Dr A.V. Spain, from an experiment reported in Spain *et al.* (1983).

Analysis Examples 5.2, 5.5, 5.9, 5.14, 5.15, 5.18, 5.19.

Set 1 : $n = 100$, Latitude $-15°43"$, Longitude $144°42"$

161	182	179	193	164	166	144	175	163	187
177	161	170	169	144	179	175	185	211	176
184	149	166	173	144	174	202	170	164	160
163	218	181	161	180	218	18	202	152	140
244	187	203	187	180	230	190	200	194	181
192	168	164	171	179	166	174	164	166	257
215	208	187	212	177	186	171	196	188	188
163	201	204	184	218	220	178	316	161	182
180	200	211	228	168	197	202	273	158	150
157	182	189	174	136	202	202	167	181	193

Set 2 : $n = 50$, Latitude $-15°32"$, Longitude $144°17"$

121	138	183	193	180	151	179	170	178	144
186	164	181	202	182	162	190	178	168	164
160	166	166	192	150	173	156	173	184	159
165	147	189	200	143	156	185	188	173	165
189	165	179	176	172	182	178	162	186	161

Set 3 : $n = 50$, Latitude $-14°59"$, Longitude $143°35"$

194	188	166	185	191	184	175	176	173	164
204	185	191	171	170	168	195	157	156	174
170	159	169	173	147	157	182	178	165	170
180	170	164	195	185	166	177	179	169	184
187	149	193	162	160	183	139	170	202	181

Set 4 : $n = 50$, Latitude $-14°19"$, Longitude $143°19"$

178	154	179	181	208	163	177	167	158	184
158	152	202	166	154	160	161	179	149	187

200	157	177	158	166	171	164	159	150	155
184	163	152	196	186	194	168	183	171	165
178	175	172	166	173	179	171	176	186	167

Set 5 : $n = 50$, Latitude −13°21", Longitude 142°53"

172	186	162	189	168	166	184	154	131	176
186	174	136	172	195	182	165	177	176	186
174	174	164	156	171	147	169	165	166	181
161	190	188	168	154	213	203	167	172	189
183	162	174	171	174	189	186	148	169	176

Set 6 : $n = 50$, Latitude −12°50", Longitude 142°44"

183	171	167	178	172	191	173	160	165	155
194	179	194	181	166	180	209	181	181	175
170	191	194	184	164	174	170	177	186	197
177	179	168	161	171	157	189	190	159	176
201	177	184	169	172	174	180	180	184	181

Set 7 : $n = 66$, Latitude −11°54", Longitude 142°30"

172	170	141	170	163	172	182	154	198	186
174	169	190	172	187	184	179	173	202	202
171	171	188	165	195	170	187	200	195	193
165	159	152	190	186	158	127	180	108	194
167	170	182	173	178	172	175	169	185	191
176	165	177	166	179	165	183	183	160	158
189	158	128	171	176	178				

Set 8 : $n = 48$, Latitude −12°06", Longitude 142°33"

166	188	192	177	114	169	251	154	198	189
178	177	189	155	165	161	175	173	182	179
178	180	190	175	161	179	156	184	187	192
168	152	173	188	175	152	166	177	161	182
179	161	187	199	140	132	178	189		

Set 9 : $n = 100$, Latitude −12°29", Longitude 142°39"

184.5	176.5	187.5	183.5	189.5	144.5	195.5	191.5	157.5	180.5
173.5	174.5	173.5	191.5	152.5	187.5	154.5	165.5	159.5	186.5
198.5	161.5	137.5	175.5	195.5	144.5	164.5	162.5	199.5	184.5
172.5	194.5	208.5	216.5	200.5	187.5	177.5	134.5	170.5	194.5

172.5	195.5	164.5	216.5	175.5	201.5	163.5	154.5	183.5	204.5
174.5	166.5	151.5	156.5	165.5	187.5	170.5	169.5	148.5	167.5
146.5	157.5	192.5	154.5	151.5	163.5	156.5	160.5	184.5	185.5
179.5	176.5	183.5	159.5	165.5	164.5	169.5	174.5	173.5	188.5
187.5	166.5	170.5	157.5	165.5	168.5	159.5	160.5	182.5	148.5
168.5	173.5	182.5	165.5	157.5	170.5	175.5	160.5	177.5	177.5

Set 10 : $n = 50$, Latitude $-13°12''$, Longitude $142°46''$

169	176	168	171	146	181	176	186	171	164
167	160	163	164	175	186	182	182	171	201
184	179	172	167	173	153	161	178	170	175
180	173	172	165	171	161	171	173	162	178
172	183	166	170	176	172	171	174	188	176

Set 11 : $n = 37$, Latitude $-15°02''$, Longitude $143°41''$

195.5	180.5	181.5	185.5	177.5	175.5	178.5	178.5	176.5	179.5
184.5	176.5	189.5	168.5	187.5	190.5	184.5	176.5	186.5	193.5
182.5	216.5	190.5	207.5	181.5	191.5	182.5	185.5	166.5	176.5
173.5	176.5	194.5	199.5	182.5	188.5	180.5			

Set 12 : $n = 31$, Latitude $-14°47''$, Longitude $143°30''$

170	167	189	227	141	178	183	180	175	160
173	168	178	179	123	191	184	180	175	188
177	189	179	179	188	183	178	176	184	155
165									

Set 13 : $n = 132$, Latitude $-13°50''$, Longitude $143°12''$

195.5	196.5	180.5	181.5	179.5	186.5	164.5	164.5	172.5	177.5
179.5	164.5	183.5	162.5	182.5	181.5	197.5	183.5	166.5	176.5
173.5	178.5	174.5	190.5	187.5	175.5	166.5	201.5	202.5	184.5
187.5	179.5	195.5	186.5	162.5	181.5	180.5	180.5	188.5	176.5
190.5	176.5	187.5	174.5	182.5	161.5	185.5	177.5	190.5	168.5
179.5	178.5	182.5	174.5	180.5	169.5	187.5	180.5	186.5	176.5
186.5	182.5	204.5	199.5	194.5	198.5	180.5	172.5	187.5	206.5
195.5	238.5	176.5	186.5	210.5	156.5	192.5	171.5	172.5	185.5
162.5	184.5	172.5	199.5	185.5	177.5	181.5	191.5	174.5	182.5
178.5	198.5	178.5	184.5	189.5	177.5	185.5	171.5	178.5	195.5
166.5	164.5	197.5	187.5	179.5	178.5	183.5	150.5	181.5	177.5
168.5	176.5	176.5	177.5	181.5	182.5	184.5	188.5	183.5	180.5

162.5 198.5 196.5 172.5 180.5 185.5 196.5 178.5 178.5 179.5
173.5 190.5

Set 14 : *n* = 92, Latitude −13°50", Longitude 143°12"

180.5	195.5	191.5	189.5	162.5	181.5	204.5	190.5	182.5	174.5
174.5	186.5	164.5	165.5	190.5	194.5	191.5	159.5	177.5	164.5
173.5	167.5	175.5	186.5	180.5	192.5	175.5	186.5	200.5	187.5
179.5	171.5	178.5	174.5	155.5	165.5	151.5	183.5	164.5	155.5
173.5	168.5	172.5	181.5	174.5	187.5	172.5	178.5	198.5	175.5
191.5	177.5	183.5	185.5	183.5	175.5	178.5	162.5	177.5	182.5
197.5	204.5	154.5	185.5	175.5	168.5	175.5	166.5	215.5	185.5
179.5	174.5	200.5	194.5	132.5	185.5	199.5	172.5	178.5	178.5
178.5	189.5	176.5	173.5	182.5	184.5	172.5	199.5	177.5	174.5
176.5	181.5								

B.14 Seasonal wind directions

Data Wind direction at Gorleston, England between 11.00am and noon on Sundays during 1968 (two values missing).

Type Vectors.

Source Mardia (1972a, Table 6.3).

Analysis Examples 5.3, 5.12.

Spring	0	20	40	60	160	170	200	220	270	290
	340	350								
Summer	10	10	20	20	30	30	40	150	150	150
	170	190	290							
Autumn	30	70	110	170	180	190	240	250	260	260
	290	350								
Winter	50	120	190	210	220	250	260	290	290	320
	320	340								

B.15 Groove and tool marks, and flute marks

Data   Groove and tool marks, and flute marks, measured in the Murruin Creek area, southwest of Yerranderie, New South Wales.

Type   Vectors (flutes) and axes (grooves and tools).

Source   Fisher & Powell (1989).

Analysis   Examples 5.4, 5.7, 5.8.

Flutes	28	354	332	59	25	43	36	51	50	48
	330	23	32	325						
Grooves	198	196	196	170	162	164	166	203	147	183
and tools	180	180	221	200	218	222	161	153	170	143
	212	224								

B.16 Cross-bed measurements from Himalayan molasse

Data   Cross-bed measurements from Himalayan molasse in Pakistan.

Type   Vectors.

Source   Wells (1990).

Analysis   Example 5.6, 5.13, 5.17.

Set 1 : 25 measurements of Rakhi Nala ripple cross-beds

10	30	30	40	50	50	60	85	90	115
135	150	150	155	170	180	190	200	205	205
230	230	235	245	250					

Set 2 : 104 measurements of Chaudan Zam large bedforms

6	7	7	21	25	32	39	50	52	58
60	65	65	68	69	70	71	72	78	80
82	87	91	93	96	96	97	97	97	99
104	104	104	106	108	108	111	115	115	115
116	117	118	118	119	120	122	125	125	130
130	130	131	132	134	138	139	142	143	143
145	146	150	150	152	154	157	159	161	162
165	166	167	167	178	180	182	185	188	192

192	195	210	217	222	222	230	240	252	266
269	271	284	285	285	285	294	299	314	318
318	330	334	350						

B.17 Orientations of rock cores

Data Orientations of core samples collected at five sampling locations in the Pacheco Pass area of the Diablo Range, California, USA.

Type Axes.

Source Upton & Fingleton (1989, Table 9.13).

Analysis Example 5.16.

Site 1	174	182	200	204	176	184	212	134		
Site 4	8	194	352	304	50	320	350	314	50	
Site 5	52	186	120	162	188					
Site 9	60	36	356	58	260	282	284	252	246	18
	358									
Site 11	210	186	208	232	242	248	180	218	222	58
	62	56	106	96	66					

B.18 Wind direction and ozone concentration

Data 19 measurements of wind direction (θ) and ozone level (x) taken at 6.00am at four-day intervals between April 18th and June 29th, 1975 at a weather station in Milwaukee.

Type Real numbers and vectors.

Source Johnson & Wehrly (1977, Table 1).

Analysis Examples 6.1, 6.4, 6.5.

θ	327	91	88	305	344	270	67	21	281	8
x	28.0	85.2	80.5	4.7	45.9	12.7	72.5	56.6	31.5	112.0
θ	204	86	333	18	57	6	11	27	84	
x	20.0	72.5	16.0	45.9	32.6	56.6	52.6	91.8	55.2	

B.19 Nest orientations and creek directions

Data   Orientations of the nests of 50 noisy scrub birds (θ) along the bank of a creek bed, together with the corresponding directions (ϕ) of creek flow at the nearest point to the nest.

Type   Vectors.

Source   Data kindly supplied by Dr Graham Smith.

Analysis Examples 6.2, 6.6.

θ	240	230	250	30	215	215	135	110	240	105
ϕ	105	75	80	105	110	75	90	100	100	80
θ	125	125	130	160	160	145	225	230	295	295
ϕ	150	135	145	130	150	125	120	140	150	140
θ	140	140	140	205	215	135	110	105	90	130
ϕ	135	150	120	135	135	130	150	150	120	150
θ	200	240	105	125	125	125	130	160	160	250
ϕ	150	130	140	180	190	190	170	180	160	185
θ	200	200	240	240	240	250	250	250	140	140
ϕ	170	180	200	190	195	180	165	170	200	175

B.20 Movements of blue periwinkles

Data   Distances (x) and directions (θ) moved by small blue periwinkles, *Nodilittorina unifasciata*, after they had been transplanted downshore from the height at which they normally live.

Type   Real numbers and vectors.

Source   Data kindly supplied by Dr A. Underwood and Ms G. Chapman.

Analysis Examples 6.3, 6.9, 6.10, 6.11.

x	107	46	33	67	122	69	43	30	12	25
θ	67	66	74	61	58	60	100	89	171	166

x	37	69	5	83	68	38	21	1	71	60
θ	98	60	197	98	86	123	165	133	101	105

x	71	71	57	53	38	70	7	48	7	21
θ	71	84	75	98	83	71	74	91	38	200

x	27
θ	56

B.21 Pairs of wind directions

Data Measurements of wind directions at 6.00am and 12.00 noon, on each of 21 consecutive days, at a weather station in Milwaukee.

Type Vectors.

Source Johnson & Wehrly (1977, Table 2).

Analysis Examples 6.7, 6.8.

θ	356	97	211	232	343	292	157	302	335	302
φ	119	162	221	259	270	29	97	292	40	313

θ	324	85	324	340	157	238	254	146	232	122
φ	94	45	47	108	221	270	119	248	270	45

θ	329
φ	23

B.22 Face-cleat in a coal seam

Data	63 measurements of median direction of face-cleat from the Wallsend Borehole Colliery, NSW, Australia.

Type   Axes.

Source   Shepherd & Fisher (1981, 1982).

Analysis Examples 7.1, 7.5.

80	90	80	84	95	79	85	80	84	90
78	90	81	90	79	90	80	84	73	115
73	85	79	72	110	75	84	75	90	85
85	70	84	96	90	73	85	90	68	124
117	114	121	127	127	125	127	84	79	73
79	79	75	80	79	68	117	119	131	122
80	90	75							

B.23 Wind directions

Data	Time series of 72 wind directions, comprising hourly measurements for three days at a site on Black Mountain, ACT, Australia.

Type   Vectors.

Source   Data kindly supplied by Dr M.A. Cameron.

Analysis Examples 7.2, 7.7, 7.8.

285	285	280	300	240	255	250	250	235	240
240	180	220	265	180	150	150	150	335	355
335	305	345	340	315	0	330	300	330	330
50	270	270	270	245	285	280	270	15	285
310	330	300	340	280	300	270	270	255	90
285	285	285	270	270	270	270	270	300	300
270	300	330	15	330	300	345	330	330	300
315	285								

B.24 Time series of flare azimuths

Data Measurements of azimuth of flare when it starts burning, based on 60 successive launches.

Type Vectors.

Source Data kindly supplied by Dr F. Lombard; description of experiment in Lombard (1986).

Analysis Example 7.6.

108.2	19.7	2.5	346.2	38.9	57.9	306.1	358.7	312.3	30.3
144.9	22.2	189.1	304.4	267.5	167.5	289.7	301.8	322.2	244.0
290.4	184.3	234.7	252.4	265.6	328.0	312.6	6.6	219.4	296.4
326.5	297.9	226.2	172.8	247.9	127.5	337.3	331.0	256.5	352.6
310.6	218.1	351.9	105.0	343.2	329.6	358.2	344.4	104.4	79.1
90.9	24.8	84.3	279.7	282.7	81.8	133.1	340.7	10.7	308.4
282.7	81.8	133.1	340.7	10.7	308.4				

References

(Numbers in parentheses at the end of each reference are page numbers for citations of that reference)

Abeyasekera, S. & Collett, D. (1982). On the estimation of the parameters of the von Mises distribution. *Commun. Statist. – Theor. Meth.* **11**, 2083–90. (92)

Abramowitz, M. & Stegun, I. A. (1970). *Handbook of Mathematical Functions.* New York: National Bureau of Standards. (50, 51)

Arsham, H. (1988). Kuiper's *P*-value as a measuring tool and decision procedure for the goodness-of-fit test. *J. Appl. Statist.* **15**, 131–5. (67)

Bagchi, P. & Guttman, I. (1990). Spuriosity and outliers in directional data. *J. Appl. Statist.* **17**, 341–50. (88)

Bagchi, P. & Kadane, J. B. (1991). Laplace approximations to posterior moments and marginal distributions on circles, spheres, and cylinders. *Canad. J. Statist.* **19**, 67–77. (93)

Barndorff-Nielsen, O. & Cox, D. R. (1979). Edgeworth and saddle-point approximations with statistical applications. (With Discussion.) *J. R. Statist. Soc.* **B 41**, 279–312. (85)

Barnett, V. & Lewis, T. (1984). *Outliers in Statistical Data.* Second edition. London: Wiley. (85)

Bartels, J. *See* Chapman *et al.* (1940).

Bartels, R. (1984). Estimation in a bidirectional mixture of von Mises distributions. *Biometrics* **40**, 777–84. (100, 102)

Batschelet, E. (1981). *Circular Statistics in Biology.* London: Academic Press. (7, 49, 50, 55, 56, 71)

Beniger, J. R. & Robyn, D. L. (1978). Quantitative graphics in Statistics: A brief history. *Am. Statist.* **32**, 1–11. (1)

Beran, R. J. (1968). Testing for uniformity on a compact homogeneous space. *J. Appl. Prob.* **5**, 177–95. (71)

Beran, R. J. (1969a). Asymptotic theory of a class of tests for uniformity of a circular distribution. *Ann. Math. Statist.* **40**, 1196–206. (30, 71)

Beran, R. J. (1969b). The derivation of nonparametric two-sample tests from tests for uniformity of a circular distribution. *Biometrika* **56**, 561–70. (123)

Beran, R. J. *See* Watson *et al.* (1967).

Best, D. J. (1979). Some easily programmed pseudo-random normal generators. *Aust. Comput. J.* **11**, 60–2. (48)

Best, D. J. & Fisher, N. I. (1979). Efficient simulation of the von Mises distribution. *Appl. Statist.* **28**, 152–57. (49)

Best, D. J. & Fisher, N. I. (1981). The bias of the maximum likelihood estimators of the von Mises–Fisher concentration parameters. *Commun. Statist. – Simul. Comput.* **B10(5)**, 493–502. (51, 92)

Boneva, L. I., Kendall, D. G. & Stefanov, I. (1971). Spline transformations: Three new diagnostic aids for the statistical data-analyst. (With Discussion.) *J. R. Statist. Soc.* **B 33**, 1–70. (30)

Breckling, J. (1989). *The Analysis of Directional Time Series: Applications to Wind Speed and Direction.* Berlin: Springer-Verlag. (189)

Cabrera, J., Schmidt-Koenig, K. & Watson, G. S. (1991). The statistical analysis of circular data. Chapter 10 (pp. 285–306) in *Human Understanding and Animal Awareness*, Editors P. P. G. Bateson & P. H. Klopfer. *Perspectives in Ethology Series*, Volume 9. New York: Plenum Press. (7, 93)

Cameron, M. A. (1983). The comparison of time series recorders. *Technometrics* **25**, 9–22. (171)

Chapman, M. G. (1986). Assessment of some controls in experimental transplants of intertidal gastropods. *J. Exp. Mar. Biol. Ecol.* **103**, 181–201. (134, 136)

Chapman, M. G. & Underwood, A. J. (1992). Experimental designs for analyses of movements by molluscs. *J. Moll. Stud.* In press. (134, 136)

Chapman, M. G. *See* Underwood *et al.* (1985).

Chapman, M. G. *See* Underwood *et al.* (1989).

Chapman, M. G. *See* Underwood *et al.* (1992).

Chapman, S. & Bartels, J. (1940). *Geomagnetism.* Two volumes. Oxford: Clarendon Press. (2)

Chatfield, C. (1989). *The Analysis of Time Series. An Introduction.* Fourth edition. London: Chapman & Hall. (185, 188)

Chayes, F. (1949). Statistical analysis of two-dimensional fabric diagrams. Chapter 22 in Fairbairn (1949). (10)

Cleveland, W. S. (1993). *The Elements of Graphing Data.* Second edition. New York: Van Nostrand Rheinhold. (110)

Cleveland, W. S. & Grosse, E. (1991). Computational methods for local regression. *Statistics and Computing* **1**, 47–62. (174)

Cohen, I. B. (1984). Florence Nightingale. *Sci. Am.* **250** (3), 98–107. (5)

Collett, D. (1980). Outliers in circular data. *Appl. Statist.* **29**, 50–57. (88)

Collett, D. & Lewis, T. (1981). Discriminating between the von Mises and Wrapped Normal distributions. *Aust. J. Statist.* **23**, 73–9. (55)

Collett, D. *See* Abeyasekera *et al.* (1982).

Cox, D. R. (1975). Contribution to Discussion of Mardia (1975). *J. R. Statist. Soc.* **B**, **37**, 380–1. (85)

Cox, D. R. *See* Barndorff-Nielsen *et al.* (1979).

Cox, D. R. & Lewis, P. A. W. (1966). *The Statistical Analysis of Series of Events.* London: Methuen & Co. Ltd. (239)

Cox, N. (1990). A note on John Playfair and the statistics of directional data. *Math. Geol.* **22**, 211–2. (5)

Creasey, J. W. *See* Fisher *et al.* (1985).

Dagpunar, J.S. (1990). Sampling from the von Mises distribution via a comparison of random numbers. *J. Appl. Statist.* **17**, 165–8. (49)

David, H. A. (1981). *Order Statistics.* Second edition. New York: Wiley. (73)

Ducharme, G. R. & Milasevic, P. (1987). Some asymptotic properties of the circular median. *Commun. Statist. – Theor. Meth.* **16** (3), 659–64. (92)

Durand, D. *See* Greenwood *et al.* (1955).

Efron, B. (1982). *The Jacknife, the Bootstrap and Other Resampling Plans.* SIAM (Regional Conference Series in Applied Mathematics, 38). Philadelphia. (199)

Efron, B. & Gong, G. (1983). A leisurely look at the bootstrap, the jacknife, and cross–validation. *Am. Statist.* **37**, 36–48. (199)

El-Atoum, S. A. M. *See* Mardia *et al.* (1976).

Embleton, B. J. J. *See* Fisher *et al.* (1987).

Epp, R. J., Tukey, J. W. & Watson, G. S. (1971). Testing unit vectors for correlation. *J. Geophys. Res.* **76**, 8480–3. (184)

Fairbairn, H. W. (1949). *Structural Petrology of Deformed Rocks.* Cambridge, Massachusetts: Addison-Wesley Press Inc. (258)

Fingleton, B. *See* Upton *et al.* (1989).

Fischer, G. (1926). Gefügeregelung und Granittektonik. *Neues Jahrb. Miner.* **B 54**, 95–114. (11)

Fisher, N. I. (1986). Robust comparison of dispersion for samples of direction data. *Aust. J. Statist.* **28**, 213–9; *Erratum,* **28**, 424. (132)

Fisher, N. I. (1987). Problems with the current definitions of the standard deviation of wind direction. *J. Clim. Appl. Meteorol.* **26**, 1522–9. (54, 55)

Fisher, N. I. (1989). Smoothing a sample of circular data. *J. Struc. Geol.* **11**, 775–8. (21, 26, 27)

Fisher, N. I. & Hall, P. G. (1989). Bootstrap confidence regions for directional data. *J. Am. Statist. Assoc.* **84**, 996–1002. Correction (1990), **85**, 608. (79)

Fisher, N. I. & Hall, P. G. (1990a). New statistical methods for directional data – I. Bootstrap comparison of mean directions and the fold test in palaeomagnetism. *Geophys. J. Internat.* **101**, 305–13. (128)

Fisher, N. I. & Hall, P. G. (1990b). On bootstrap hypothesis testing. *Aust. J. Statist.* **32**, 177–90. (202)

Fisher, N. I. & Hall, P. G. (1991a). Bootstrap algorithms for small samples. *J. Stat. Plan. Inference.* **27**, 157–69. (117, 203)

Fisher, N. I. & Hall, P. G. (1991b). Bootstrap methods for directional data. In *The Art of Statistical Science: A Tribute to G.S. Watson,* edited by K.V. Mardia. Chichester: Wiley. (93)

Fisher, N. I., Huntington, J. F., Jackett, D. R., Willcox, M. E. & Creasey, J. W. (1985). Spatial analysis of two-dimensional orientation data. *Math. Geol.* **17**, 177–94. (194)

Fisher, N. I. & Lee, A. J. (1981). Nonparametric measures of angular–linear association. *Biometrika* **68**, 629–36. (144, 233)

Fisher, N. I. & Lee, A. J. (1982). Nonparametric measures of angular–angular association. *Biometrika* **69**, 315–21. (151, 235)

Fisher, N. I. & Lee, A. J. (1983). A correlation coefficient for circular data. *Biometrika.* **70**, 327–32. (57, 151, 153, 237)

Fisher, N. I. & Lee, A. J. (1986). Correlation coefficients for random variables on a unit sphere or hypersphere. *Biometrika.* **73**, 159–64. (153)

Fisher, N. I. & Lee, A. J. (1992). Regression models for an angular response. *Biometrics.* **48**, 665–77. (155, 158, 164, 167, 168)

Fisher, N. I. & Lee, A. J. (1993). Time series analysis of circular data. *J.R. Statist. Soc.* **B**. To appear. (186, 187, 189)

Fisher, N. I. & Lewis, T. (1983). Estimating the common mean direction of several circular or spherical distributions with differing dispersions. *Biometrika* **70**, 333–41, *Errata,* **71** (1984), 655. (79, 89, 92, 121)

Fisher, N. I., Lewis, T. & Embleton, B. J. J. (1987). *Statistical Analysis of Spherical Data.* Cambridge: Cambridge University Press. (xv, 14, 85)

Fisher, N. I. & Powell, C. McA. (1989). Statistical analysis of two-dimensional palaeocurrent data: Methods and examples. *Aust. J. Earth Sci.* **36**, 91–107. (77, 79, 114, 121, 226, 242, 250)

Fisher, N. I. *See* Best *et al.* (1979).

Fisher, N. I. *See* Best *et al.* (1981).

Fisher, N. I. *See* Shepherd *et al.* (1981).

Fisher, N. I. *See* Shepherd *et al.* (1982).

Freedman, L. S. (1979). The use of a Kolmogorov–Smirnov type statistic in testing hypothesis about seasonal variation. *J. Epidem. Comm. Health* **33**, 223–28. (71)

Freedman, L. S. (1981). Watson's U_N^2 statistic for a discrete distribution. *Biometrika* **68**, 708–11. (71)

Funkhouser, H. G. (1936). A note on a Tenth Century Graph. *Osiris* **1**, 260–2. (1, 3)

Gadsden, R. J. & Kanji, G. K. (1981). Sequential analysis for angular data. *Statistician* **30**, 119–29. (103)

Gadsden, R. J. & Kanji, G. K. (1982). Sequential analysis applied to circular data. *Sequential Anal.* **1**, 305–14. (103)

Gadsden, R. J. *See* Harrison *et al.* (1986).

Gong, G. (1983). *See* Efron *et al.* (1982).

Goodyear, C. P. (1970). Terrestrial and aquatic orientation in the Starhead Topminnow, *Fundulus notti. Science* **168**, 603–5. (241)

Gould, A. L. (1969). A regression technique for angular variates. *Biometrics* **25**, 683–700. (168)

Gower, J. C. (1966). Some distance properties of latent rest and vector methods used in multivariate analysis. *Biometrika* **53**, 325–38. (197)

Graedel, T. E. (1977). The wind boxplot: an improved wind rose. *J. Appl. Meteorol.* **16**, 448–51. (112)

Graves, T. S. (1979). *Randomization Procedures for a Two Way Analysis of Orientation Data.* Ph.D. thesis, Cornell University, Ithaca NY. (134)

Greenwood, J. A. & Durand, D. (1955). The distribution of length and components of the sum of *n* random unit vectors. *Ann. Math. Statist.* **26**, 233–46. (71)

Grosse, E. *See* Cleveland *et al.* (1991).

Gumbel, E. J. (1954). Applications of the circular normal distribution. *J. Am. Statist. Assoc.* **49**, 267–97. (13)

Guttman, I. *See* Bagchi *et al.* (1990).

Hall, P. G. (1992). *The Bootstrap and Edgeworth Expansion.* New York: Springer. (199, 203)

Hall, P. G. *See* Fisher *et al.* (1989).

Hall, P. G. *See* Fisher *et al.* (1990a).

Hall, P.G. *See* Fisher *et al.* (1990b).

Hall, P. G. *See* Fisher *et al.* (1991a).

Hall, P. G. *See* Fisher *et al.* (1991b).

Halley, E. (1701). *The Description and Uses of a New Seachart of the Whole World, Shewing Variations of the Compass.* London. (2, 4)

Hansen, K. M. & Mount, V. F. (1990). Smoothing and extrapolation of coastal stress orientation measurements. *J. Geophys. Res.* **95 B2**, 1155–65. (170, 192)

Härdle, W. (1990). *Applied Nonparametric Regression.* Cambridge: Cambridge University Press. (173)

Hartley, H. O. *See* Pearson *et al.* (1970).

Harrison, D., Kanji, G. K. & Gadsden, R. J. (1986). Analysis of variance for circular data. *J. Appl. Statist.* **13**, 123–38. (134)

Harrison, D. & Kanji, G. K. (1988). The development of analysis of variance for circular data. *J. Appl. Statist.* **15**, 197–223. (134)

Hawkins, D. M. (1980). *Identification of Outliers.* London: Chapman & Hall. (85)

Herbert, T. J. (1983). The influence of axial rotation upon interception of direct solar radiation by plant leaves. *J. Theor. Biol.* **105**, 603–18. (174)

Herbert, T. J. & Larsen, P. B. (1985). Leaf movement in *Calathea lutea* (*Marantaceae*). *Oecologia* **67**, 238–43. (174)

Hermans, M. & Rasson, J. P. (1985). A new Sobolev test for uniformity on the circle. *Biometrika* **72**, 698–702. (71)

Hill, G. W. (1976). New approximations to the von Mises distribution. *Biometrika* **63**, 673–676. *Erratum*, (1977a), **64**, 655. (52)

Hill, G. W. (1977). Incomplete Bessel function I_0; the von Mises distribution [S14]. *ACM Trans. Math. Softw.* 3. (52)

Hill, I. D. *See* Wichman *et al.* (1982).

Hollander, M. & Wolfe, D. A. (1973). *Nonparametric Statistical Models.* New York: Wiley. (229)

Hsu, Y.-S., Walker, J. J. & Ogren, D. E. (1986). A step-wise method for determining the number of component distributions in a mixture. *Math. Geol.* **18**, 153–60. (101)

Huntington, J. F. *See* Fisher *et al.* 1985.

IMSL (1991). *The IMSL Libraries.* Edition 2.0. Sugar Land, Texas: IMSL Inc. (44, 49, 51, 97)

Jackett, D. R. *See* Fisher *et al.* (1985).

James, W. R. *See* Jones *et al.* (1969).

Jander, R. (1957). Die optische Richtangsorientierung der roten Waldameise (*Formica rufa. L.*). *Z. vergl. Physiologie* **40**, 162–238. (243)

Jeffreys, H. (1961). *Theory of Probability.* Third edition. Oxford: Oxford University Press. (45)

John, R. D. *See* Spain *et al.* (1983).

Johnson, R. A. & Wehrly, T. E. (1977). Measures and models for angular correlation and angular–linear regression. *J. R. Statist. Soc.* **B 39**, 222–9. (57, 251, 253)

Johnson, R. A. & Wehrly, T. E. (1978). Some angular–linear distributions and related regression models. *J. Am. Statist. Assoc.* **73**, 602–6. (168)

Johnson, R. A. *See* Wehrly *et al.* (1979).

Jones, T. A. & James, W. R. (1969). Analysis of bimodal orientation data. *Math. Geol.* **1**, 129–35. (102)

Jupp, P. E. & Mardia, K. V. (1980). A general correlation coefficient for directional data and related regression problems. *Biometrika* **67**, 163–73. (57)

Jupp, P. E. & Mardia, K. V. (1989). A unified view of the theory of directional statistics, 1975–1989. *Int. Statist. Rev.* **57**, 261–94. (14, 57, 153)

Kadane, J. B. *See* Bagchi *et al.* (1991).

Kanji, G. K. *See* Gadsden *et al.* (1981).

Kanji, G. K. *See* Gadsden *et al.* (1982).

Kanji, G. K. *See* Harrison *et al.* (1986).

Kanji, G. K. *See* Harrison *et al.* (1988).

Kendall, D. G. *See* Boneva *et al.* (1971).

Kent, J. T. & Tyler, D. E. (1988). Maximum likelihood estimation for the wrapped Cauchy distribution. *J. Appl. Statist.* **15**, 247–54. (46, 103)

Kiepenheuer, J. (1978). Pigeon homing: A repetition of the deflector loft experiment. *Behav. Ecol. Sociobiol.* **3**, 393–95. (7)

Kinderman, A. J. & Monahan, J. F. (1977). Computer generation of random variables using the ratio of uniform deviates. *ACM Trans. Maths. Software* **3**, 257–60. (48)

Klotz, J. (1964). Small-sample power of the bivariate sign tests of Blumen and Hodges. *Ann. Math. Statist.* **35**, 1576–82. (56)

Koutbeiy, M. A. *See* Spurr *et al.* (1991).

Krumbein, W. C. (1939). Preferred orientations of pebbles in sedimentary deposits. *J. Geol.* **47**, 673–706. (11, 243)

Kuiper, N. H. (1960). Tests concerning random points on the circle. *Koninklijke Nederlandse Akademie Van Wetenschappen, Proceedings Series A* **LXIII**, 38–47. (71)

Larsen, P. B. *See* Herbert *et al.* (1985).

Lawson, A. B. (1991). GLIM and normalising constant models in spatial and directional analysis. *Comput. Stat. Data Anal.* To appear. (85)

Laycock, P. J. (1975). Optimal regression: regression models for directions. *Biometrika* **62**, 305–11. (168)

Lee, A. J. *See* Fisher *et al.* (1981).

Lee, A. J. *See* Fisher *et al.* (1982).

Lee, A. J. *See* Fisher *et al.* (1983).

Lee, A. J. *See* Fisher *et al.* (1986).

Lee, A. J. *See* Fisher *et al.* (1992).

Lee, A. J. *See* Fisher *et al.* (1993).

Lehmacher, W. & Lienert, G. A. (1980). Note on a binomial test against sectoral preference of circular observations. *Biom. J.* **22**, 249–52. (71)

Lehmann, E. L. (1975). *Nonparametrics: Statistical Methods based on Ranks.* San Francisco: Holden-Day, Inc. (81)

Lenth, R. (1981). Robust methods of location for directional data. *Technometrics* **23**, 77–81. (92)

Lewis, P. A. W. *See* Cox, *et al.* (1966).

Lewis, T. *See* Collett *et al.* (1981).

Lewis, T. *See* Barnett *et al.* (1984).

Lewis, T. *See* Fisher *et al.* (1983).

Lewis, T. *See* Fisher *et al.* (1987).

Liddell, I. G. & Ord, J. K. (1978). Linear–circular correlation coefficients: some further results. *Biometrika* **65**, 448–50. (145)

Lienert, G. A. *See* Lehmacher *et al.* (1980).

Lockhart, R. A. & Stephens, M. A. (1985). Tests of fit for the von Mises distribution. *Biometrika* **72**, 647–52. (85, 230)

Lombard, F. (1986). The change-point problem for angular data: a nonparametric approach. *Technometrics* **28**, 391–97. (182, 183, 255)

Lombard, F. (1988). Detecting change points by Fourier analysis. *Technometrics* **30**, 305–10. (183)

MacKinnon, W. J. (1964). Table for both the sign test and distribution-free confidence intervals of the median for sample sizes to 1000. *J. Am. Statist. Assoc.* **59**, 935–56. (73)

McLeod, A. I. (1985). A remark on AS183. An efficient and portable pseudo-random number generator. *Appl. Statist.* **34**, 198–200. (44)

Mardia, K. V. (1972a). *Statistics of Directional Data.* London: Academic Press. (14, 35, 41, 45, 49, 50, 51, 55, 56, 71, 93, 102, 124, 128, 243, 249)

Mardia, K. V. (1972b). A multi-sample uniform scores test on a circle and its parametric competitor. *J. R. Statist. Soc.* **B 34**, 102–13. (123)

Mardia, K. V. (1975). Statistics of directional data. (With Discussion.) *J. R. Statist. Soc.* **B 37**, 349–93. (49, 151)

Mardia, K. V. (1976). Linear–circular correlation coefficients and rhythmometry. *Biometrika* **63**, 403–5. (141, 145, 231)

Mardia, K. V. & El-Atoum, S. A. M. (1976). Bayesian inference for the von Mises–Fisher distribution. *Biometrika* **63**, 203–6. (93, 129, 131)

Mardia, K. V. & Spurr, B. D. (1973). Multisample tests for multimodal and axial circular populations. *J. R. Statist. Soc.* **B 35**, 422–36. (123, 128)

Mardia, K. V. & Sutton, T. W. (1975). On the modes of a mixture of two von Mises distributions. *Biometrika* **62**, 699–701. (102)

Mardia, K. V. *See* Jupp *et al.* (1980).

Mardia, K. V. *See* Jupp *et al.* (1989).

Marriot, F. H. C. (1969). Associated directions. *Biometrics* **25**, 775–6. (154)

Marriott, F. H. C. (1990). *A Dictionary of Statistical Terms*, Fifth edition. Harlow: Longman Scientific & Technical. (14)

Mendoza, C. E. (1986). Smoothing unit vector fields. *J. Math. Geol. 18* 307–22. (192)

Milasevic, P. *See* Ducharme *et al.* (1987).

Mitchell, J. (1767). An inquiry into the probable parallax and magnitude of the fixed stars, from the quantity of light which they afford us, and the particular circumstances of their situation. *Philos. Trans. R. Soc.* **57**, 423–33. (2)

Mount, V. F. *See* Hansen *et al.* (1990).

Monahan, J. F. *See* Kinderman *et al.* (1977).

Moore, B. R. (1980). A modification of the Rayleigh test for vector data. *Biometrika* **67** 175–80. (71)

Müller, M. *See* Wehner *et al.* (1985).

NAG (1991). *NAG Fortran Library. Release 5.* Oxford: Numerical Algorithms Group. (97)

Nightingale, F. (1858). *Notes on Matters Affecting the Health, Efficiency, and Hospital Administration of the British Army.* London: Harrison & Sons. (5, 6, 10)

Nijenhuis, A. & Wilf, H. S. (1978). *Combinatorial Algorithms for Computers and Calculators.* Second edition. New York: Academic Press. (143, 144, 147, 202, 215, 216, 217)

Ogren, D. E. *See* Hsu *et al.* (1986).

Okello-Oloya, T. *See* Spain *et al.* (1983).

Ord, J. K. *See* Liddell *et al.* (1978).

Pabst, B. & Vicentini, H. (1978). Dislocation experiments in the migrating seastar *Astropecten jonstoni. Marine Biology* **48**, 271–8. (87, 245)

Pearson, E. S. & Hartley, H. O. (1970). *Biometrika Tables for Statisticians, Volume I.* Third edition. Cambridge: Cambridge University Press. (126, 132)

Pettijohn, F. J. *See* Potter *et al.* (1977).

Pfeifer, M. A. (1985). *See* Weinberg *et al.* (1984).

Playfair, J. (1802). *Illustrations of the Huttonian Theory of the Earth.* London: Cadwell & Davies. Facsimile reprint: New York, Dover (1964). (5)

Potter, P. E. (1955). The petrology and origin of the Lafayette gravel. Part 1. Mineralogy and petrology. *J. Geol.* **63**, 1–38. (12)

Potter, P. E. & Pettijohn, F. J. (1977). *Paleocurrents and Basin Analysis.* Second edition. Berlin: Springer-Verlag. (12, 18)

Powell, C. McA. *See* Fisher *et al.* (1989).

Rao, J. S. (1969). Some variants of chi-square for testing uniformity on the circle. *Z. Wahrscheinlichkeitstheor. verwandte Geb.* **22**, 33–44. (71)

Rao, J. S. (1984). Nonparametric methods in directional data. Chapter 31 in *Handbook of Statistics, Volume 4. Nonparametric Statistics*, editors P. R. Krishnaiah & P. K. Sen, pp. 755–70. Amsterdam: Elsevier Science Publishers. (71, 123)

Rao, J. S. & Yoon, Y. (1983). Comparison of the limiting efficiencies of two chi-square type tests for the circle. *J. Indian Statist. Assoc.* **21**, 19–26. (85)

Rasson, J. P. *See* Hermans *et al.* (1985).

Rayleigh, *Lord* (1880). On the resultant of a large number of vibrations of the same pitch and of arbitrary phase. *Phil. Mag.* **10**, 73–8. (10, 71)

Rayleigh, *Lord* (1905). The problem of random walk. *Nature* **72**, 318. (10, 71)

Rayleigh, *Lord* (1919). On the problem of random vibrations, and of random flights in one, two or three dimensions. *Phil. Mag.* **37**, 321–47. (10, 71)

Reiche, P. (1938). An analysis of cross-lamination. The Coconino Sandstone. *J. Geol.* **46**, 905–32. (11, 13)

Rivest, L. P. (1982). Some statistical methods for bivariate circular data. *J. R. Statist. Soc.* **B 44**, 81–90. (57)

Robyn, D. L. *See* Beniger *et al.* (1978).

Rothman, E. D. (1971). Tests of coordinate independence for a bivariate sample on a torus. *Ann. Math. Statist.* **42**, 1962–9. (154)

Salvemini, T. (1940). On a mean trigonometric function, in the case of cyclic series. *Atti del secondo Congresso dell' Unione Matematica Italiana*, Bologna April 4–6, 1940. Reprinted in *Scritti Scelti* (1981), C.I.S.V. a r.l.–Roma. (13)

Salvemini, T. (1942). On the mean trigonometric function. *Atti della V Riunione della Società Italiana di Statistica.* Rome May 30–31 & June 1, 1942. Reprinted in *Scritta Scelti* (1981), C.I.S.V. a r.l.–Roma. (13)

Schach, S. (1969). Nonparametric symmetry tests for circular distributions. *Biometrika* **56**, 571–7. (81)

Schmidt, W. (1917). Statistische Methoden beim Gefügestudium Kristalliner Schiefer. *Sitz. Kaiserl. Akad. Wiss. Wien, Math. – nat. Kl. Abt.* 1, **126**, 515–38. (10, 11, 12, 48)

Schmidt-Koenig, K. (1963). On the role of the loft, the distance and site of release in pigeon homing (the "cross-loft experiment"). *Biol. Bull.* **125**, 154–64. (245)

Schmidt-Koenig, K. *See* Cabrera *et al.* (1991).

Schou, G. (1978). Estimation of the concentration parameter in the von Mises–Fisher distributions. *Biometrika* **65**, 369–77. (92)

Scott, D. W. & Sheather, S. J. (1985). Kernel density estimation with binned data. *Commun. Statist. – Theor. Meth.* **14**, 1353–9. (27)

Sheather, S. J. *See* Scott *et al.* (1985).

Shepherd, J. & Fisher, N. I. (1981). A rapid method of mapping fracture trends in collieries. *Aust. Coal Miner* August, 24–33. (183, 254)

Shepherd, J. & Fisher, N. I. (1982). Rapid method of mapping fracture trends in collieries. *Trans. Soc. Min. Eng. AIME* **270**, 1931–2. (183, 254)

Silverman, B. W. (1986). *Density Estimation for Statistics and Data Analysis.* London: Chapman and Hall. (24, 26, 27, 30)

Smith, N. M. (1988) *Reconstruction of the Tertiary drainage systems of the Inverell region.* Unpublished B.Sc.(Hons.) thesis, Department of Geography, University of Sydney, Australia. (240, 242)

Spain, A. V., Okello-Oloya, T. & John, R. D. (1983). Orientation of the termitaria of two species of *Amitermes* (Isoptera:Termitinae) from Northern Queensland. *Aust. J. Zoo.* **31**, 167–77. (246)

Spurr, B. D. & Koutbeiy, M. A. (1991). A comparison of various methods for estimating the parameters in mixtures of von Mises distributions. *Commun. Statist. – Simul. Comput.* **20** 725–41. (98, 102)

Spurr, B. D. *See* Mardia *et al.* (1973).

Stefanov, I. *See* Boneva *et al.* (1971).

Stegun, I. A. *See* Abramowitz *et al.* (1970).

Steinmetz, R. (1962). Analysis of vectorial data. *J. Sediment. Petrol.* **32**, 801–12. (12)

Stephens, M. A. (1965). The goodness-of-fit statistic V_N: distribution and significance points. *Biometrika* **52**, 309–21. (71)

Stephens, M. A. (1969a). Tests for the von Mises distribution. *Biometrika* **56**, 149–60. (93, 96, 128)

Stephens, M. A. (1969b). Techniques for directional data. Technical Report #150, Dept. of Statistics, Stanford University, Stanford, CA. (23, 102, 241)

Stephens, M. A. (1970). Use of the Kolmogorov–Smirnov, Cramer–von Mises and related statistics without the use of tables. *J. R. Statist. Soc.* **B 32**, 115–22. (71)

Stephens, M. A. (1974). EDF tests for goodness of fit and some comparisons. *J. Amer. Statist. Assoc.* **69**, 730–7. (225)

Stephens, M. A. (1982). Use of the von Mises distribution to analyse continuous proportions. *Biometrika* **62**, 197–203. (133)

Stephens, M. A. *See* Lockhart *et al.* (1985).

Strasser, S. *See* Wehner *et al.* (1985).

Sutton, T. W. *See* Mardia *et al.* (1975).

Tukey, J. W. (1977). *Exploratory Data Analysis.* Addison-Wesley Publishing Company: Reading, MA. (110)

Tukey, J. W. *See* Epp *et al.* (1971).

Tyler, D. E. (1987). Statistical analysis for the angular central Gaussian distribution on the sphere. *Biometrika* **74**, 579–89. (46)

Tyler, D. E. *See* Kent *et al.* (1988).

Underwood, A. J. & Chapman, M. G. (1985). Multifactorial analyses of directions of movement of animals. *J. Exp. Mar. Biol. Ecol.* **91**, 17–43. (134, 136)

Underwood, A. J. & Chapman, M. G. (1989). Experimental analyses of the influences of topography of the substratum on movements and density of an intertidal snail, *Littorina unifasciata. J. Exp. Mar. Biol. Ecol.* **134**, 175–96. (134, 136)

Underwood, A. J. & Chapman, M. G. (1992). Experiments on topographic influences on density and dispersion of *Littorina unifasciata* in New South Wales. *J. Moll. Stud.* In press. (134, 136)

Underwood, A. J. *See* Chapman *et al.* (1990).

Upton, G. J. G. (1973). Single-sample tests for the von Mises distribution. *Biometrika* **60**, 87–99. (96)

Upton, G. J. G. (1974). New approximations to the distribution of certain angular statistics. *Biometrika* **61**, 369–73. (93)

Upton, G. J. G. (1986). Approximate confidence intervals for the mean direction of a von Mises distribution. *Biometrika* **73**, 525–7. (93)

Upton, G. J. G. & Fingleton, B. (1989). *Spatial Data Analysis by Example. Volume 2. Categorical and Directional Data.* New York: John Wiley. (7, 71, 87, 93, 123, 128, 132, 245, 251)

Vicentini, H. *See* Pabst *et al.* (1978).

von Mises, R. (1918). Über die "Ganzzahligkeit" der Atomgewichte und Verwandte Fragen *Physikal. Z.* **19**, 490–500. (11)

Walker, J. J. *See* Hsu *et al.* (1986).

Watson, G. S. (1961). Goodness-of-fit tests on the circle. *Biometrika* **48**, 109–14. (71, 85)

Watson, G. S. (1966). The statistics of orientation data. *J. Geol.* **74**, 786–97. (8)

Watson, G. S. (1967). Some problems in the statistics of directions. *Bull. Int. Statist. Inst.* (36th Session of the ISI, Sydney Australia), **42**, 374–85. (30)

Watson, G. S. (1969). Density estimation by orthogonal series. *Ann. Math. Statist.* **40**, 1469–98. (30)

Watson, G. S. (1970). Orientation statistics in the earth sciences. *Bull. geol. Instn Univ. Uppsala* N.S. **2**:9, 73–89. (8)

Watson, G. S. (1971). *Selected Topics in Statistical Theory.* Notes on lectures given at the 1971 MAA Summer Seminar, Williams College, Williamstown, Massachusetts: Mathematical Association of America. (30)

Watson, G. S. (1983). *Statistics on Spheres.* New York: Wiley. (50, 51, 56, 79, 92, 117, 121, 128, 131)

Watson, G. S. (1985). Interpolation and smoothing of directed and undirected line data. In *Multivariate Analysis – VI*, edited by P. R. Krishnaiah, pp. 613–25. New York: Academic Press. (175, 190)

Watson, G. S. & Beran, R. J. (1967). Testing a sequence of unit vectors for randomness. *J. Geophys. Res.* **72**, 5655–9. (184)

Watson, G. S. & Williams, E. J. (1956). On the construction of significance tests on the circle and the sphere. *Biometrika* **43**, 344–52. (14, 56, 92, 128)

Watson, G. S. *See* Cabrera *et al.* (1991).

Watson, G. S. *See* Epp *et al.* (1971).

Watson, G. S. *See* Wheeler *et al.* (1964).

Wehner, R. (1982). Himmelsnavigation bei Insekten. Neurophysiologie und Verhalten. *Neujahrsblatt der Naturforschenden und Gesellschaft Zürich* **184**, 1–132. (7)

Wehner, R. & Müller, M. (1985). Does interocular transfer occur in visual navigation by ants? *Nature* **315**, 228–9. (244)

Wehner, R. & Strasser, S. (1985). The POL area of the honey bee's eye: behavioural evidence. *Physiol. Entomol.* **10**, 337–49. (244)

Wehrly, T. E & Johnson, R. A. (1979). Bivariate models for dependence of angular observations and a related Markov process. *Biometrika* **66**, 255–6. (57)

Wehrly, T. E. *See* Johnson *et al.* (1977).

Wehrly, T. E. *See* Johnson *et al.* (1978).

Weinberg, C. R. & Pfeifer, M. A. (1984). An improved method for measuring heart-rate variability: assessment of cardiac autonomic function. *Biometrics* **40**, 855–61. (82)

Wells, N. A. (1990). Comparing sets of circular orientations by modified chi-squared testing. *Comput. Geosci.* **16**, 1155–70. (250)

Wheeler, S. & Watson, G. S. (1964). A distribution-free two-sample test on the circle. *Biometrika* **51**, 256–7. (123)

Wichman, B. A. & Hill, I. D. (1982). Algorithm AS183: An efficient and portable pseudo–random number generator. *Appl. Statist.* **31**, 188–90. (44)

Wilf, H. S. *See* Nijenhuis *et al.* (1978).

Willcox, M. E. *See* Fisher *et al.* (1985).

Williams, E. J. *See* Watson *et al.* (1956).

Yoon, Y. *See* Rao *et al.* (1983).

Young, D. S. (1987a). Random vectors and spatial analysis by Geostatistics for geotechnical applications. *Math. Geol.* **19**, 467–79. (193)

Young, D. S. (1987b). Indicator kriging for unit vectors: rock joint orientations. *Math. Geol.* **19**, 467–79. (193)

Zar, J. H. (1984). *Biostatistical Analysis.* Second edition. Englewood Cliffs, NJ: Prentice-Hall Inc. (7)

Index